Nuclear Asymmetry and Deterrence

This book offers a broader theory of nuclear deterrence and examines the way nuclear and conventional deterrence interact with non-military factors in a series of historical case studies.

The existing body of literature largely leans toward the analytical primacy of nuclear deterrence and it is often implicitly assumed that nuclear weapons are so important that, when they are present, other factors need not be studied. This book addresses this omission. It develops a research framework that incorporates the military aspects of deterrence, both nuclear and conventional, together with various perceptual factors, international circumstances, domestic politics, and norms. This framework is then used to re-examine five historical crises that brought two nuclear countries to the brink of war: the hostile asymmetric nuclear relations between the United States and China in the early 1960s; between the Soviet Union and China in the late 1960s; between Israel and Iraq in 1977–1981; between the United States and North Korea in 1992–1994; and, finally, between the United States and the Soviet Union during the 1962 Cuban Missile Crisis. The main empirical findings challenge the common expectation that the threat of nuclear retaliation represents the ultimate deterrent. In fact, it can be said, with a high degree of confidence, that it was rather the threat of conventional retaliation that acted as a major stabilizer.

This book will be of much interest to students of nuclear proliferation, Cold War studies, deterrence theory, security studies and IR in general.

Jan Ludvik is Assistant Professor at the Department of Security Studies and Research Fellow at the Center for Security Policy, Charles University, Czech Republic. He has a PhD in International Relations.

Routledge Global Security Studies
Series editors: Aaron Karp and Regina Karp

Global Security Studies emphasizes broad forces reshaping global security and the dilemmas facing decision-makers the world over. The series stresses issues relevant in many countries and regions, accessible to broad professional and academic audiences as well as to students, and enduring through explicit theoretical foundations.

Nuclear Asymmetry and Deterrence

Theory, policy and history

Jan Ludvik

Routledge
Taylor & Francis Group

LONDON AND NEW YORK

First published 2017
by Routledge

2 Park Square, Milton Park, Abingdon, Oxfordshire OX14 4RN
52 Vanderbilt Avenue, New York, NY 10017

Routledge is an imprint of the Taylor & Francis Group, an informa business

First issued in paperback 2020

British Library Cataloguing-in-Publication Data
A catalogue record for this book is available from the British Library

Library of Congress Cataloging-in-Publication Data
Names: Ludvik, Jan.
Title: Nuclear asymmetry and deterrence : theory, policy and history /
Jan Ludvik.
Description: Abingdon, Oxon ; New York, NY : Routledge is an
imprint of the Taylor & Francis Group, an Informa Business, [2017] |
Series: Routledge global security studies | Includes bibliographical
references and index.
Identifiers: LCCN 2016021372| ISBN 9781138696198 (hardback) |
ISBN 9781315525174 (ebook)
Subjects: LCSH: Deterrence (Strategy) | Nuclear weapons. | World
politics–20th century. | Deterrence (Strategy)–Case studies. | Nuclear
weapons–Case studies. | World politics–20th century–Case studies.
Classification: LCC U162.6 .L83 2017 | DDC 355.02/17–dc23
LC record available at https://lccn.loc.gov/2016021372

ISBN: 978-1-138-69619-8 (hbk)
ISBN: 978-0-367-66805-1 (pbk)

Typeset in Times New Roman
by Wearset Ltd, Boldon, Tyne and Wear

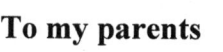

To my parents

Contents

Tables

Preface

Years ago, as a fresh and enthusiastic doctoral student, I started the research that culminated in this book with the simple goal to understand how nuclear deterrence works between a state with a small nuclear arsenal and its more powerful enemy. I was convinced that nuclear deterrence cannot be as easy as most people think. Well acquainted with the work of Cold War deterrence theorists and strategists, I was struggling with the puzzle. Was this considerable group of people completely wrong? Excellent human minds had tried hard to figure out how much is enough to deter. Consequently they designed large nuclear postures because they did not believe that small numbers could suffice. Of course, they could have been completely wrong. Already during the Cold War, some dissenting opinions suggested that very little is enough. And most importantly, the cases of clear deterrence failures between two nuclear states, even with small arsenals, are missing. Something must have helped stability to prevail. Going through the empirical evidence, I started to sense that, in fact, there really is a critical omission. Nuclear deterrence was given analytical primacy to the extent that few scholars seriously considered that conventional deterrence works in nuclear dyads as well. Yet, it seems that conventional deterrence worked better in a number of cases. Limiting research to studying how nuclear deterrence works between big and small cannot be enough.

For a good student of deterrence, a mere stable coexistence of nuclear weapons and stability in a hostile dyad should not imply nuclear deterrence success. Of course, nuclear deterrence is a plausible explanation, but not the only one. Deterrence is a complex phenomenon which is remarkably difficult to study empirically. Its failures, at least, are observable, easy-to-recognize events. Yet clear-cut deterrence failures are scarce in nuclear dyads. Deterrence successes, on the other hand, are essentially non-events. When deterrence holds, the war is avoided. But often it is hard to say that it was avoided and even more difficult to say why. It is well recognized that equifinality is a critical problem in the social sciences in general, and it is an even more challenging problem in the study of deterrence successes. It is widely recognized that it should not be underestimated in deterrence scholarship. But more often than not, this advice is ignored or only paid lip service in appropriate methodological sections.

Other plausible causes should be taken into account for control, particularly where nuclear deterrence operates with a small and vulnerable nuclear arsenal. First, a state with a small nuclear arsenal is certainly armed with conventional weapons as well. It can use them not only to defend itself, but also to retaliate. Researchers usually fail to pay sufficient attention to this. Second, numerous domestic, international, or normative factors may in fact affect the challengers' decisions. In most empirical cases, more than one factor operates simultaneously in a delicate interplay. All this should be taken into account. Therefore, I had to broaden the original scope of my research to pay adequate attention to the complexity of deterrence – the way nuclear and conventional deterrence interact with non-military factors in the dyads between a state with a small nuclear arsenal and its more powerful enemy. This brings revealing results. The main empirical findings challenge the common expectation that the threat of *nuclear retaliation* represents the ultimate deterrent. In fact, it can be said, with a high degree of confidence, that, in the empirical cases studied in this book, where deterrence succeeded it was rather the threat of *conventional retaliation* that acted as a major stabilizer.

These findings have important implications for both theory and policy. But I would never have been able to get there without help from many people. I am heavily indebted to them – repaying this debt may not be even possible.

First and foremost, I would like to acknowledge the contribution of my friend and former supervisor Nik Hynek. This book would have never been written without his encouragement and occasional friendly-minded coercion. We spent hours discussing this research and Nik's comments were critical for the development of my ideas and my research approach. I cannot be more grateful for his help.

Michal Smetana was our counterpart in a number of discussions, and his comments were no less useful. He deserves the warmest thanks. So does Luděk Moravec, with whom we have almost institutionalized the practice of discussing, reading, and commenting on each other's research. Nikolai Sokov and Jan Eichler carefully read the manuscript despite their tight schedule, and their comments unquestionably helped me further improve the text. My colleagues from the Center for Security Policy at the Charles University in Prague were a supportive collective, and without this base, it would have been very difficult to finish writing.

Andrew Humphrys enthusiastically accepted the proposal and sent the manuscript out for the review, tolerating my daring claim that I intentionally wrote my thesis as a book therefore it does not need much rewriting. He and Hannah Ferguson led me through the dark alleys of the publishing process.

Furthermore, Routledge commissioned two anonymous reviewers – their comments were excellent suggestions for revisions.

I am also happy to acknowledge the support from The Czech Science Foundation's grant project GA16–02288S.

Last but not least, my family's support was indispensable. My wife Jana survived all the ups and downs that accompanied my research. She kept me in touch with the real world and I owe her and our little kids a lot of time that I spent with the book instead of them.

Needless to say, all errors are entirely my own.

Introduction

This is a book about the way nuclear and conventional deterrence interact with non-military factors in dyads involving a state with a small nuclear arsenal and its more powerful enemy. Contrary to a large body of deterrence literature, this book is primarily interested in small nuclear arsenals rather than big ones, or, more accurately, in relations between the two. Trying to move current deterrence theory further, it is asking a critical question: how does deterrence work between a country with a small nuclear arsenal and a more powerful challenger and what makes it work? Why another book about deterrence when previously published volumes already occupy a considerable space on the bookshelves? Most importantly, because there is a practical need for a broader theory of deterrence. While the existing body of literature includes many excellent books, it largely leans toward the analytical primacy of nuclear deterrence. It is often implicitly assumed that nuclear weapons are so important that, when they are present, other factors need not be studied. This is empirically unwarranted, in particular with small nuclear arsenals.

The subject of this study has great importance for both theory and policy. As a matter of fact, the proliferation of nuclear weapons is not only recognized as one of the most worrying threats to international security, but also as one that is likely to continue in the years to come. It is hard to imagine that newcomers into the nuclear club will be able to skip the period of smallness. So far, every nuclear-armed country has had to live through such a period. Yet newcomers are also likely to be in somewhat hostile relations with one or another existing power. Apparently, at this point, it is necessary to ask, should the world be concerned? Will such asymmetric relations be stable? At least since the early Cold War, when the NATO stepped in that direction – mostly to save budgets from expensive conventional armament – nuclear weapons earned the reputation of great equalizers that could offset conventional weakness. Yet is it really so, particularly with small nuclear forces? The most recent crisis in Ukraine is likely to spark this debate again. More than a decade ago, John Mearsheimer recommended the young

Ukrainian state to keep its nuclear weapons to be able to live next to the much stronger Russia.[1] Now Ukraine is hinting it may reconsider its decision to give up nuclear weapons. But is going nuclear likely to protect the country from Russian aggression or is it rather likely to invite it?

The theoretical value of the subject matter is no less important. Voluminous theoretical literature has been written about deterrence since the heydays of strategic studies in the 1950s. Yet this literature is in many respects remarkably incomplete. First, the empirical and theoretical legs of deterrence literature are off-balance. Historically, the major body of literature was theoretical, rational choice driven, and prescriptive. It is widely recognized that more empirical studies would be useful to enrich this scholarship. Second, deterrence literature had for decades paid attention predominantly to the world's most important nuclear deterrence dyad, namely the one between the United States and its NATO allies, on one side, and the Soviet Union and its Warsaw Pact allies, on the other. While such a bias is understandable and not unwarranted, it left the issues of small arsenals rather under-researched. The third, and perhaps the greatest, omission in the state of the art of deterrence literature is the aforementioned analytical primacy of nuclear deterrence.

In an attempt to bridge this gap and move current deterrence theory further, this book develops a rich research framework that incorporates the military aspects of deterrence, both nuclear and conventional, together with various perceptual factors, international circumstances, domestic politics, and norms. This procedure is inspired by the method of comparative case studies that was pioneered in deterrence literature by Alexander George.[2] It is primarily inductive and empirical, rather than deductive and theoretical, yet this does not mean it is atheoretical. In this sense it is not purely inductive as it starts with a range of theoretical concepts to structure its focus and allows systematic cross-case comparison. This framework is then used to re-examine five historical crises that brought two nuclear countries to the brink of war. Four of them represent deterrence successes, while one is a deterrence failure. Four cover crises between an emerging or recent proliferator and an established nuclear power. Furthermore, one crisis between the Cold War superpowers, where the relations of power were much less asymmetrical, is included for control.

The structure of this book is as follows. Chapter 1 explains the rationale behind the need for a broader theory of deterrence, the book's research strategy and the conceptual framework in which this study is situated. First, it introduces the vibrant scholarly debates about nuclear deterrence, about the role rationality plays therein, and the stabilizing/destabilizing effects of nuclear proliferation on international relations. It also outlines the brief and, regrettably, incomplete debate about conventional deterrence. Building on this, the chapter identifies several omissions and biases in the available literature that are relevant for this book, and outlines how these challenges are to be addressed. It argues that the weaknesses of various deterrence theories prevent this book from explicitly building on a single predecessor. Theoretical plurality is suggested as a solution, and a focused set of concepts is identified to guide this study. The concepts subsequently serve as a basis for the procedure of structured focused comparison. Chapter 1 then proceeds with a more thorough explanation of book's case selection and the basic rules for cross-case comparison.

The five subsequent chapters are devoted to empirical cases covering the history of hostile relations between the United States and China in the early 1960s; the Soviet Union and China in the late 1960s; Israel and Iraq in 1977–1981; the United States and North Korea in 1992–1994; and the United States and the Soviet Union in 1962. All these cases are severe crises, general deterrence failures, where deterrence mechanisms should be best observable. All the case studies are similarly structured. The first part sketches a historical narrative of the particular crisis. The context is outlined together with available details of the challenger's planning and his sensitivities to the deterrer's threats. The second part of each empirical chapter is devoted to a brief description of the case's details related to the military aspects of deterrence, perceptional factors, international circumstances, domestic politics, and the role of norms, i.e., the concepts which structure focused comparison in this study. Chapter 2 outlines the debates about the option of destroying China's emerging nuclear program that took place in the subsequent administrations of President Kennedy and President Johnson. Chapter 3 introduces the escalation of the Sino-Soviet split in 1969 that reportedly lead the Kremlin to contemplate solving the Chinese problem "once and for all." Chapter 4, the single case of deterrence failure, reviews the details of Israeli destruction of Iraq's Osiraq nuclear reactor. Chapter 5 explores the Clinton administration's responses to North Korea's 1994 nuclear challenge, particularly with respect to U.S. debates of the "Osiraq option." Finally, Chapter 6 presents a control case, reviewing the details of the Cuban Missile crisis with regard to the structure of this study. As a case where the deterrer operated a fairly large nuclear arsenal, it incorporates a vital control allowing cross-case comparison of the previous small-arsenal cases.

Chapter 7 subsequently provides the synthesis of this research. Drawing on the empirical evidence, it first challenges the analytical primacy of nuclear deterrence. It shows a revealing pattern of nuclear deterrence failure which accompanies deterrence with small arsenals unless the deterrer is able to pass the threshold of the second strike. Following this, the chapter highlights the importance of conventional military threats for stabilizing small-to-big dyads in the small arsenal's period of vulnerability. In this respect, it ascribes a decisive role to the largely omitted threat of conventional retaliation. Apart from these two major findings, the synthetic chapter also formulates several tentative findings which deserve the further attention of future researchers.

Notes

1 John J. Merasheimer, "The Case for a Ukrainian Nuclear Deterrent," *Foreign Affairs*, 72/3 (Summer 1993) 50–66.
2 Alexander L. George and Richard Smoke, *Deterrence in American Foreign Policy* (New York: Columbia University Press, 1970).

1 Toward a broader theory of deterrence

Enormous volumes have been written about deterrence since the emergence of strategic studies as the field accessible not only to military professionals but also to civilian researchers. It is no coincidence that the emergence of strategic studies in academia overlaps with the advance of nuclear weapons in the real world, and conscious study of deterrence as a major aspect of international politics in the second half of the 1940s. Lack of previous experience with nuclear weapons was a formative factor for the early study of deterrence. Military services lost their previous almost unchallengeable monopoly on the study of strategy. Unable to claim established expertise in the field that was new to everyone, soldiers had to admit civilians into their jealously guarded sphere. But the civilians had no better empirical experience than the military. What allegedly made a civilian expert superior to his military counterpart was the ability to work in a field that, lacking empirical evidence, started as almost purely theoretical.[1]

Existing theories and the need of theoretical plurality

In those formative years, scholars from the so-called "first cycle of strategic studies" mostly addressed the practical problems related to American deterrence of the Soviet Union such as basing patterns of long-range strategic bombers.[2] Yet its practical orientation did not prevent the development of concepts that would play a central role in the study of deterrence in the years to come. By a coincidence, Albert Wohlstetter and his colleagues in RAND discovered the concept of second strike.[3] Glenn Snyder theorized the distinction between deterrence by denial and deterrence by punishment.[4] Thomas Schelling coined the term "compellence" as a persuasive compatriot of dissuasive deterrence in the field of coercive strategies.[5] Those are only a few examples. Considering the lack of empirical data, the progress made by early students of deterrence is as impressive as it is impossible to do justice to all the influential scholars in this brief review.[6]

At the beginning, scholarship was mostly guided by researchers' intuition, qualified guessing, and illustrative examples. Nonetheless, this lack of scientific rigor soon came under sharp criticism. As a response, deterrence scholars increasingly turned to rigorous deductive methodologies, most importantly to game theory.

Drawing on several powerful simplifying assumptions, deductive rational deterrence theory established itself as a leading scientific product in deterrence scholarship.[7] The principal assumptions taken by rational deterrence scholars generally include the rational actor assumption, the state as unitary actor assumption, and explanation of variation in outcomes by differences in actor opportunities.[8] Logically coherent, rigorous, and parsimonious, rational deterrence theory has clear advantages. Yet it fails to grasp the complexity of deterrence in the real world. In fact, its heuristic value is limited at best. Using comparative case studies and advancements in the study of cognitive psychology, various scholars successfully highlighted the deficiencies of the dominant rational deterrence paradigm.[9] Up until the end of the Cold War, the utility of rational deterrence theory was intensively discussed as it came under sharp criticism. But lacking an adequate theoretical replacement, its critics never really dethroned rational deterrence theory.[10] The sudden lack of interest in the further study of deterrence that accompanied the end of the bipolar nuclear rivalry then left the debate open and incomplete, with unfinished results.

It is no surprise that during the formative years of the Cold War the study of nuclear weapons affairs mostly revolved around the crucial U.S.–USSR nexus. Correspondingly, the attention paid to small nuclear arsenals was limited. Initiatives in the theoretical study of the topic were mostly left to the nationals of emerging nuclear states, most importantly to French strategists.[11] Their ideas often met with skepticism in mainstream strategic studies.[12] Yet, the intellectual underpinnings of the French school of deterrence helped establish the existential deterrence theory, whose proponents were led by former national security advisor to President Kennedy and President Johnson, McGeorge Bundy.[13] These existentialists, mostly following the rational deterrence theory dictum and assuming that it is irrational to risk nuclear destruction of even one of one's own cities, hold that deterrence rests with the uncertainty that is always present in a nuclear relationship, even with small numbers. Accordingly, even a very high probability of a successful first strike against a very irresolute defender does not allow ruling out the threat of retaliation. This theorizing of the existentialists had only limited influence on the U.S.–Soviet deterrence modality. Nevertheless, its principal ideas are well reflected in the key debate about the effectiveness of deterrence with small arsenals, and particularly in what Vipin Narang calls existential bias.[14]

This noteworthy debate took place between so-called proliferation optimists and proliferation pessimists and is yet to be resolved. Its origins can be traced to Kenneth Waltz's influential and controversial contribution about the effects of nuclear proliferation on international stability.[15] Waltz, drawing from his well-known neorealism and rational deterrence theory, argues powerfully that the general negative image of nuclear weapons proliferation is incorrect. In his view, deterrence is easy to achieve with nuclear weapons and almost certain to hold as no challenger will risk the threat of nuclear retaliation, however unlikely it may be. Waltz's approach, soon labeled as proliferation optimism, does not lack support.[16] Nevertheless, proliferation optimism remains the view of an influential minority, and has since been challenged by numerous more pessimistic scholars.[17] The pessimists try to illuminate various deficiencies of this logic. Among them is the underestimation of the threat of preventive war in the early stage of nuclear arsenal development;[18] the overestimation of the level of nuclear safety that can be easily reached by new nuclear states;[19] the unreasonably low threshold for the ability to reach invulnerable second-strike forces;[20] the necessity to include the theory of nuclear operations into the predominantly structural realist logic;[21] or the likely increase of the threat of nuclear terrorism.[22] Others have highlighted that the positions of both optimists and pessimists are equally incomplete[23] and all sides generally agree that more thorough empirical research is necessary to confirm or disprove the presented arguments.

It is remarkable that this call to underpin the optimist-pessimist debate with empirical evidence is valid even in the debate's key part devoted to proliferation's impact on international stability. So far the evidence has been largely introduced in an illustrative fashion. Pessimists argue that proliferation undermines stability since preventive strikes were discussed almost from the time when it was recognized that nuclear weapons were a feasible option. Even before the United States successfully tested the first bomb in 1945, the allies targeted German facilities to prevent Nazis from its acquisition.[24] After the war, prominent leaders in the United States advocated preventive strikes against the Soviet Union. In the 1960s, first the United States and then the Soviet Union considered preventive options that would have taken out the Chinese nuclear program. In 1981, the Israeli air force destroyed Iraq's Osiraq nuclear plant, then under construction. In 1994 a serious crisis arose between United States and North Korea over the DPRK's nuclear weapons program and Washington was once again getting ready for a preventive strike. This evidence cannot be easily ruled out. However, the optimist position seems at least equally strong. Nuclear weapons were neither used, nor subject to attack, in any of the aforementioned cases. Whenever a strike took place, the targeted program had not reached its final phase. Thus optimists hold that prevention against a nuclear-armed country is close to impossible. The attacker cannot ever be sure of complete success. With some underlying possibility of nuclear retaliation always present, nuclear deterrence should make the world more stable. Systematic comparative historical examination is missing from the debate. The evidence tends to be limited as the authors draw predominantly from two cases: the U.S.–Soviet Cold War experience and the Indo-Pakistani nuclear rivalry.[25] Even though both cases are important ones, it can be argued that thorough exploration has been avoided of cases even more critical.[26]

Lyle Goldstein's *Preventive Attack and Weapons of Mass Destruction* is a notable exception. Goldstein thoroughly examined five cases of nuclear dyads between hostile states: the United States and the Soviet Union in the early Cold War, the United States and China in the 1960s, the Soviet Union and China in the 1960s, China and India in the 1970s and 1980s, and Israel and Iraq in the 1970s and 1980s. To improve the validity of his research, he also added seven mini cases from other periods of the aforementioned hostile dyads, of U.S. relations with rogues in the 1990s, and of India and Pakistan in the 1990s. Goldstein – drawing on this substantial empirical dataset – argues that "radically asymmetric nuclear (WMD) relationships are fundamentally unstable, because nascent arsenals do not deter effectively," and that a "given superior power's propensity to attack the weaker state's nascent nuclear (WMD) arsenal will substantially depend on the conventional balance, alliance dynamics, norms, and geography."[27] Goldstein's contribution is particularly valuable because he challenged the strongest claim of the optimists camp, namely that nuclear deterrence stabilized nuclear dyads because no war involving nuclear weapons ever happened. Most pessimists systematically examine how nuclear deterrence may fail when bureaucratic, psychological, organizational and other biases undermine its effectiveness. But Goldstein persuasively shows that nuclear deterrence might not effectively work at all in some hostile nuclear dyads and suggests that other factors, particularly conventional deterrence, better explain the lack of war between nuclear powers in radically asymmetric rivalries. Yet Goldstein's research can benefit from further refinement. All the five cases he fully develops cover deterrence by small arsenals, yet for a more valid comparison, his research can be improved using a control case of deterrence with a more advanced nuclear weapons complex. Furthermore, his comparative framework is somewhat underspecified, and particularly the way conventional and nuclear deterrence overlap, enhance, or replace each other needs more clarification.

This task is as needed in deterrence scholarship as it is demanding. A common fallacy in security studies is to treat nuclear and conventional deterrence separately. Presumably, nuclear weapons have qualities that make the study of conventional deterrence in nuclear dyads irrelevant. Conventional deterrence is allegedly prone to miscalculation.[28] With conventional weapons, leaders are more likely to believe that the cost of a conflict will be acceptable. Even when the costs turn out to be much higher in protracted conventional conflicts, the actors have time to adjust and accept losses that would have been prohibitive at the time when hostilities started.[29] Thus the effect of conventional deterrence compared to nuclear deterrence is almost nil. In dyads where both conventional and nuclear threats operate, the study of conventional factors must yield.

Therefore, in accordance to the general preoccupation of scholarship with nuclear warfare, relatively little attention was devoted to a thorough study of regular war and conventional deterrence from the end of the Second World War at least till the 1970s.[30] While conventional deterrence was pursued regularly by the Cold War contenders throughout 1950s and 1960s, its systematic research treatment lagged behind. Only then did conventional deterrence emerge more prominently in the literature as it seemed that Soviet tanks might grab large portions of Western Europe under the stability-instability paradox.[31] However, a major limiting factor remained despite the unquestionable value of this scholarship: conventional threats were treated separately from nuclear ones. The interplay of conventional and nuclear deterrence was rather put aside. It was assumed that under the stability-instability paradox, nuclear threats keep the other side's nuclear forces at bay, whereas a limited conventional conflict is plausible. Of major interest was to study how conventional weapons deter a "small" regional conflict when a "big" global one is made impossible by nuclear arsenals. Yet conventional deterrence does not operate only in purely conventional dyads or in dyads where nuclear deterrence creates the stability-instability paradox, making nuclear deterrence inoperative for certain smaller contingencies.

Having read this brief sketch of the available scholarship, the reader should not be left in doubt. The progress in deterrence scholarship since the late 1940s has been truly impressive. Something that started almost from the scratch developed over the years into some of the most respectable research programs in strategic studies. A number of the most brilliant human minds have contributed to the field. Yet, for historical reasons often related to the practical need to advise on those days' most pressing security issues, other issues were put aside and shortcuts were sometimes necessary. Strategic studies literature on nuclear deterrence is thus, in some respects, notably incomplete. The first cycle did invaluable exploratory work, yet methodologically it rarely crossed the threshold of qualified estimates. Rigorous and logically consistent, yet often failing to pass empirical tests, rational deterrence theory traded validity for rigor and parsimony. Proliferation optimists opened an important debate by applying the rational deterrence dictum on the allegedly destabilizing proliferation that creates new small arsenals, but did little to repair the theory's flawed construct. Critics of rational deterrence, armed with tools from political psychology's arsenal and utilizing comparative case studies methodology, made an important step forward. Nevertheless, little attention has been paid to the interplay of conventional and nuclear deterrence. And students of conventional deterrence repeated the same mistake by limiting their studies to the realm where nuclear deterrence was inoperative. This flaw needs to be repaired.

Yet the problem appears clearer than the solution. The study of nuclear deterrence poses a special challenge to the social scientist. Deterrence successes are often events of enormous historical importance. However, they are not easy to observe. As Patrick Morgan noted, "when an attack does not occur, how do we know why?" In a sense, studying deterrence means studying non-events.[32] When deterrence works, the status quo prevails. But when neither party challenges the status quo, the status quo prevails as well. Good students of deterrence literally cannot over-appreciate the problem of equifinality.[33] In the case of nuclear weapons the researcher's troubles are exacerbated by a uniform dependent variable. It is a matter of fact that, by now, the world has witnessed eight decades of life with nuclear weapons and it is likely that some more will pass till the noble goal of nuclear weapons elimination, "the global zero," can be achieved.[34] Luckily for the world, but less luckily for deterrence students, nuclear weapons were used only twice at the beginning of the nuclear age – against the Japanese cities of Hiroshima and Nagasaki. And, as the bombs were used at the end of a bloody conventional war against an almost broken enemy that had no chance to retaliate in a similar fashion or escalate the conflict further, the two cases can shed little light on the study of deterrence. Students of nuclear deterrence thus lack an unquestionable, clear-cut example of deterrence failure. With a uniform dependent variable – no nuclear war – it is all too easy to assume that the possession of nuclear weapons on both sides is the key explanatory variable for peace in nuclear dyads.

A broader theory of deterrence is needed to address the aforementioned challenges of analytical primacy of nuclear deterrence and of equifinality. But as with any theory building it cannot avoid the tradeoff between richness and parsimony.[35] Here I take the first approach. I am inspired by a plurality of theories and I trade parsimony for richness. The choice is not unwarranted. The way of omitting conventional and, to a lesser extent non-military, factors in the study of nuclear dyads does not allow starting with one theory; none of them is satisfactory enough, yet most of them can enrich the research. It also requires illuminating the role conventional and non-nuclear factors play in nuclear dyads. Here, one possibility, the more parsimonious one, would be to build three separate models – nuclear, conventional, and non-nuclear – and test them against the same empirical cases. Scott Sagan used a similar approach in his seminal, often cited study of the causes of nuclear proliferation.[36] Nonetheless, this strategy cannot properly grasp a part of the picture where the causes overlap and interact. It is still possible to argue, in a similar manner, with Sagan that exclusive models do not preclude multiple causes in a single case. But the complexity of the possible interplay between nuclear, conventional, and non-military factors would be lost.

Building a conceptual framework

The complexity of the situation does not preclude this book from starting with a simple model of deterrence. On the one side of deterrence stands the expected utility of the challenger. What can the challenger gain from starting military action and how valuable is it to him? The other side of the coin shows the expected costs. The deterrer actively manipulates this cost perception by the threat of military counteraction. The challenger estimates the magnitude of this threat and its credibility, which in simple language means the likelihood that the threat will be fulfilled. But the costs are obviously seldom limited to the military moves of the deterrer. States are not two billiard balls in the middle of nowhere. Even the staunchest realist must admit that other balls/states are moving on the table and may add costs into the challenger's equation. Foreign policy analysts will quickly add that domestic factors must be considered and constructivists will add the role of norms and institutions. Nonetheless, the threat of military counteraction is essential in deterrence. Should the non-military costs suffice to preserve peace, the situation cannot be treated as a case of deterrence otherwise the category will be unreasonably broadened.

Even though the military component is a necessary part of deterrence, non-military factors must be part of an analysis. Non-military costs add to military costs in dissuading the challenger in a delicate interplay. To be able to grasp the complexity of nuclear, conventional, and non-nuclear factors, this book is armed with a population of major concepts from a wide diversity of existing literature about nuclear and conventional deterrence and international conflicts. Using available data from major crises between a small nuclear state and a formidable challenger, I later access the role various concepts play in this delicate deterrence game in a cross-case comparative framework. I am not giving priority to any theory or concept. Yet the introduction of concepts must start somewhere, and parsimonious rational deterrence theory is a good choice. Most of its concepts are recognized as valid by other works, which usually add what rational deterrence theory allegedly omits.

The essence of deterrence, not only for rational deterrence theory, is the deterrer's threat of military response to the challenger's action. Though it is possible the deterrer is bluffing, the possession of military hardware is a vital consideration. For nuclear deterrence, this means the deterrer's nuclear arsenal, which can be described in absolute, relative, or threshold-based terms. What matters more for deterrence cannot be yet established from the empirically underdeveloped literature. Existentialists argue that asymmetry does not matter;[37] proponents of minimum deterrence add that some threshold (usually second-strike ability) must be passed;[38] supporters of MAD argue that stability requires symmetry in the ability to substantially hurt the other side;[39] proponents of war-fighting agree that an all-out strategic war can be deterred under symmetry, but call for nuclear primacy to allow for nuclear weapons to serve in other contingencies.[40] I am using three concepts to grasp these options. *Deterrer's nuclear arsenal* describes absolute numbers, *nuclear asymmetry* relative ones, and *second-strike criterion* the threshold most commonly acknowledged in the literature.

By deterrer's nuclear arsenal I mean the key physical attributes of the respective nuclear complexes: weapon types, numbers, delivery vehicles, the command, control, and communication systems. Due to the design of this research, the number will always be rather small. Nonetheless, significant variety may exist in smallness. Not only can it vary from an uncertain zero to several dozens, but a great variety is possible in terms of the entire complexes of delivery vehicles, concealment strategies, early warning systems, communications, and other aspects. Those attributes can now usually be estimated with a high degree of confidence from available sources.

Nuclear asymmetry describes the relative ratio between challenger and deterrer. Strong nuclear asymmetry means a situation when a substantially smaller (quantitatively) and less sophisticated nuclear force (qualitatively) faces a qualitatively and quantitatively larger nuclear force. Should the two complexes be on par in one category, the asymmetry will be considered limited, while equality does not need explaining.

The second-strike criterion – the ability to retaliate after the attacker's first strike – is probably the most common criterion of stable nuclear deterrence. Optimists consider second strike to be the only strategic requirement for the stabilizing effects of nuclear deterrence.[41] Yet they put the threshold for passing the second-strike criterion unreasonably low. Waltz argues that "if the country attacked has even a rudimentary nuclear capability, one's own severe punishment becomes possible."[42] Nuclear weapons are allegedly small and easy to hide and can be delivered through many means, including non-traditional ones such as trucks driven from neighboring countries or boats lying offshore.[43] Any state that has nuclear weapons will pass such relaxed requirements. Yet to treat the concepts in a similar fashion has little analytical value. Thus I argue that it is necessary to return to an earlier understanding of the second strike. Going back to the original meaning introduced by Wohlstetter, I understand passing the second strike threshold as a force that has the ability to survive enemy attack, make and communicate the decision to retaliate, overcome the enemy's active defense, and destroy a valuable target despite its passive defense.[44]

Similarly to the nuclear realm, conventional deterrence needs military hardware to back up the threat. The simplest assumption about conventional deterrence is that it will hold where defense is stronger than offense. This is a good starting point yet by itself it says little. First, defense may be stronger on a general scale, or on a theater one. Second, an explanation should be added as to what makes offense stronger then defense. The easiest explanation is that the stronger side is the one with bigger numbers. Other common explanations build on the state of the technology in either systemic or dyadic terms. More sophisticated models also discuss the role of force employment and training.[45] None of the approaches is perfect. The less sophisticated models usually perform in parsimony, but fail terribly when confronted with empirical evidence.[46] It is clear that numbers must be taken into account, as well as the technology of both sides, the skills of their forces, and their employment strategies. Estimating preponderance is thus an uneasy task. Fortunately, my cases are less demanding in terms of estimating balance. I will consider the answer satisfactory when estimating the clear preponderance of either challenger or deterrer, or treating the situation as a rough equality. By *general conventional preponderance* I mean the situation when one side's armed forces are, in general, substantially stronger in terms of numbers, technology, training, and employment strategies, while by *theater conventional preponderance* I mean the situation when one side's armed forces are substantially stronger in terms of numbers, technology, training, and employment strategies on a theater where targets valuable to the challenger can be found.

Both nuclear and conventional deterrence are influenced by the state of technology. Whilst the state of technology is already incorporated in the aforementioned concepts, the influential literature on the revolution in military affairs (RMA) calls for independent treatment of the *challenger's technological advantage*.[47] It can be described as a significant advantage of the challenger's major weapons systems that would be employed in the case of conflict in terms of state-of-the-art sophistication over the deterrer's weapons systems. The impact of the latest technological changes, often known under the RMA label, on deterrence is under-researched yet the implications are possibly profound.[48] Current technology, particularly improvements in weapons accuracy, allows replacing nuclear weapons by conventional weapons against most point targets such as the enemy's nuclear complexes, command, control, and communication systems.[49] Similarly, satellite imagery with continuously improving resolution makes locating the targets easier, even though recent wars such as NATO's bombardment of Kosovo and the American invasion of Iraq highlighted the limits of modern technologies.[50] However, there is little new in the interest in the impact of technological change on deterrence. Advancements in technology such as the introduction of ICBMs, terminal guidance improvements, MIRVs, or contemplated deployments of missile defense are often believed to significantly undermine deterrence. They have subsequently led to the necessity of improvements in target hardening and the introduction of mobile weapons platforms, which should have restored the offence-defense balance.

A purely military component of deterrence also needs specific *availability of information*. Even the most dreadful weapon systems cannot influence the challenger's calculations unless he has some indications of their existence. Perfectly secret weapons do not deter. Yet informing the challenger about a weapon's existence is likely to attract his attention and make him search for more details, locations, ways to destroy or paralyze it. Taking this route undermines deterrence. To what extent it does so remains the function of the challenger's ability to get information accurate enough to make the first strike an attractive option. The challenger needs knowledge about the targets' locations and defensive systems, which is established from sources that the challenger deems credible. Optimists argue that with nuclear weapons it is close to impossible. They are small, easily hidden and easily moved from one location to another. Even if the challenger can learn about the existence of, location of, and way to destroy all existing weapons, he will not be certain that he really knows about all of them.[51] However, this may not necessarily be critical. Most importantly, optimists disregard the role of the strategic initiative which will belong to the challenger. Strategic surprise has played a critical role in a number of conflicts.[52] The challenger may only need critical information available for a brief period of time to strike. With the help of modern technologies, human intelligence, insider information, and other elements, the optimists may be too optimistic about the impossibility of knowing enough and for certain.

The military means that allow fulfilling the threat are one side of the coin. The willingness to do it is the other. At this point, deterrence theory speaks about credibility. Allegedly, credibility was not a problem in the pre-nuclear age. The stronger side could carry out its threat because it had the means. The weaker side could not because it could not overcome the enemy's troops on the battlefield. Then, as a game changer, came the threat of nuclear retaliation, allowing for robust punishment without a realistic defense.[53] Before the nuclear revolution (and the advancement of strategic bombing) the stronger side would pay only the direct costs of its military activity on the battlefield, but the nation's homeland, population, and industry behind the front line could remain intact. With the costs of conflict rising enormously for both sides, neither of them can rationally fulfil its military threat because both would lose more than they can gain. Thomas Schelling devised a solution to this critical problem for rational deterrence theory, or more accurately a way around.[54] Schelling argues that, in nuclear coercion, actors can credibly threaten or take steps on the route that may eventually result in the situation getting out of control.[55] The more resolute state will be ready to go closer to the point when the situation gets out of hand, it will take more risks and thus it will prevail in the crisis. In the world of rational deterrence theory, the more resolute state is the one that has a higher interest at stake.

However, without complete information, the challenger only knows the importance of the interests at stake to him and must estimate the importance that the deterrer attaches to his. As a matter of fact, states seldom fight for what they do not consider very important. This was true even in the age of limited wars among European kings, long before nationalism followed by the industrial revolution made war total and before the nuclear revolution exacerbated it further. Yet in spite of the apparent costs, states often fight for things that do not appear worthy of fighting for to outsiders. Asking how reasonable it appears to the observer is the wrong question. What matters is how important even an ephemeral thing appears to the party that contemplates whether to go to war. By *centrality of conflict* I mean not only the absolute importance of the issue at stake, but also the relative importance of the conflict dyad between the challenger and the deterrer to other existing conflict dyads where the challenger is a party, from the challenger's subjective perspective. Similarly, the challenger must estimate the deterrer's commitment to fighting should the deterrence fail, which is determined by the importance of the same issues at stake for the deterrer. If this *perceived resolve of the deterrer* were low he would be unlikely to go to war as he would be viewed as irresolute.

The psychology school in deterrence scholarship argues that this simple model is unsatisfactory. Decision making is influenced by biases that do not correspond with rational deterrence theory and also rational deterrence theory says little about threat perception,[56] which critically depends on previous experience with others' behavior.[57] Communication between actors is essential, yet it is also the place where the bulk of information can be lost or misunderstood.[58] Misunderstanding is more likely between actors who lack established communication channels and a history of mutual relations. Therefore, Darryl Howlett suggests differentiating between three ideal types of deterrence relationships: established, semi-established, and non-established. The *institutionalization of mutual relations* describes the degree of shared expectations about the requirements of stable deterrence and the existence of proven formal or informal communication channels. Howlett argues that established relationships that are characterized by high institutionalization, either formal or informal, generate a reasonable degree of expectation about the future behavior of the other side and a common understanding about the requirements of stable deterrence. This useful mutual understanding is much lower in semi-established relationships, and almost missing in non-established relationships. In semi-established relationships, institutionalization already exists in one form or another yet the actors have not learned the rules that are accepted by both parties; in non-established relationships, actors lack historical and procedural interaction concerning the meaning of stability.[59]

A track record of conflict in the dyad that shapes the understanding of the other side's intentions can be a vital condition of institutionalization. The learning curve essential for a shared understanding of deterrence can hardly emerge in non-hostile dyads; deterrence does not work between friends but between adversaries. Yet this is not a sufficient condition. For instance, for the 40 years of the Cold War, the United States was on hostile terms with both the Soviet Union and North Korea. But whereas highly institutionalized relations, regular contacts, red lines, and an understanding of what the other side's military moves usually meant developed with Moscow, almost no contacts were held between Washington and Pyongyang. The difference may have profound implications. While a high degree of institutionalization will likely underpin deterrence and reduce misunderstanding, a strong *history of hostility* may have a different effect. The other side's signals are more likely to be viewed as threatening, and conciliatory gestures can be viewed as a sign of weakness. Where hostility is not balanced by a better understanding of the other side, the game is open to escalation and deterrence failure.

Failure is even more likely when leaders believe that whatever the risks are now, they will be bigger in the future. The situation may appear as a *last resort*. The deterrer can threaten unacceptable damage, have all the necessary means to carry out his threat, and communicate the threat with sufficient credibility, but deterrence can still fail when the challenger sees the situation as the last chance to prevent an even more unacceptable development. This last resort thinking is even possible when in reality acceptable ways out of crises exist and are only hidden. Fortunately, preparations for a war take time and previously unconsidered solutions or re-evaluations of the situation often appear. But this does not mean that deterrence worked at that moment; in fact it failed for a moment that was not long enough to start the shooting.

In the nuclear proliferation realm, last-resort situations will often correspond with the *perception of the deterrer's rationality*. The logic behind this link is simple. Living next to a nuclear-armed hostile nation appears to be uncomfortable but as long as the nation is rational, peaceful coexistence under mutual deterrence is possible. Yet should the other side be irrational, prone to hazardous behavior, and even possibly ready to use its nuclear weapons despite the likely consequences, then the coexistence only turns into waiting for a disaster. Recently, the administration of George W. Bush raised this issue to prominence in public discourse.[60] Washington alleged that the rogue states designated by them were so reckless that they would use nuclear weapons in spite of the threat of severe punishment. The same deterrence strategy that kept the Soviets (whose rationality was questioned during the Cold War yet apparently improved in the American eyes 20 years later) at bay could not work against Saddam Hussein, Kim Yong-Il, and their like.[61] A serious stability problem is embodied in this assumption. Notwithstanding the real preferences of the rogue state, if someone is believed to be reckless, aggressive, irrational, nuclear-armed and essentially undeterrable, the only options are to be ready to defend against them, or to strike preventively and eliminate the threat. The imperfections of defense against nuclear weapons then make the preventive strike even more tempting.

Three more international factors cannot be safely omitted from my study of deterrence. First, simple *third-party deterrence* means the threat of involvement into the original conflict by a third party. The deterrer's allies are likely candidates for taking or threatening actions that decisively influence the challenger's decision, a move that is known in strategic studies as extended deterrence.[62] Alternatively, a third party that is not aligned to either side may threaten to align with the deterrer, or withdraw its support for the challenger should the deterrence fail.[63]

The factor of *alliance politics* is the next option. The challenger's allies may vocally or tacitly disagree with an action that is contemplated by the challenger. In international relations, where treaties are not enforced by any higher authority, alliances depend on trust between allies. Countries that depend on alliances will find it unattractive to hurt their ties for anything less than vital. In fact, allies may not even tacitly express their opposition and yet the challenger may be self-deterred by assuming that the move would hurt their alliance.

International action legitimacy is the third international constraint on the challenger's behavior. Contrary to the aforementioned concepts that are largely based on actors' self-interests and others' abilities to enforce, the effects of international action legitimacy are based on an actor's sense that a certain rule should be obeyed.[64] While the former two are more inspired by realist and liberal literature, the latter bears much from constructivism. State sovereignty, the deeply internalized norm of nonintervention into another nation's affairs, mitigates the state's willingness to intervene. Yet this does not make peace certain. First, some nations do not accept the legitimacy of the entire system or a part thereof. Second, the normative expectations about the legitimacy of military intervention into another country's affairs change over time.[65] Thus interventions into another country's affairs are increasingly perceived as legitimate, particularly in cases where other international norms such as human rights or anti-proliferation are being breached.[66]

One more constructivist concept, the normative prohibition on the use of nuclear weapons, has sparked a thorough debate in strategic studies and greatly contributed to the development of constructivism itself. Nina Tannenwald who introduced the concept of *nuclear taboo* argues that "its effect has been to delegitimize nuclear weapons as weapons of war."[67] This strong normative prohibition has two plausible effects on a challenger's decision making. The stronger effect delegitimizes the strike against an enemy's nuclear arsenal altogether, as nuclear weapons would be involved in war, though only as passive targets. The weaker effect limits the challenger's military options by ruling out his nuclear weapons as suitable weapons for a strike, yet keeps nuclear weapons as legitimate targets.

Most of the aforementioned concepts are not very useful for studying the role domestic politics play in decision making about international affairs. For the sake of theory, actors tend to be treated as "like units." Of course, this is not to say domestic politics is always ignored, yet it mostly lies outside the scope of the various theories that inspired the concepts covered on the previous pages. To be able to enrich this analysis by incorporating the domestic dimension in the research design, I am utilizing concepts inspired by liberalism and the foreign policy analysis subfield of international relations which, in the words of Valerie Hudson, offers an "actor-specific theory as the ground of international relations."[68]

Domestic political considerations unquestionably belong to plausible explanations of non-intervention. Leaders in democratic countries must carefully watch for *domestic action legitimacy* among their domestic audiences, should they seek re-election. Leaders in non-democratic countries may find their hands less tied, yet even most non-democratic regimes cannot fully ignore the lack of domestic support for their military action. Also, irrespective of people's preferences, opposition may appear among the top leaders themselves. Should this happen in a democratic coalition government, the risk of breakup may appear too costly to the hawks. Non-coalitions will likely find it easier to overcome internal opposition, but again even the strongest dictators need the support of their powerbase, and therefore, possible *opposition among influential decision makers* cannot be disregarded. And as different considerations may appear across different regime types, the *regime type* should be considered as well in this research, with democratic peace theory as the best example of scholarship.[69] Though the theory is still debated and its mechanisms put under criticism,[70] it is little surprising that none of my dyads has a democratic challenger and a democratic deterrer.

This by no means small number of concepts should cover the critical factors that available literature, unquestionably valuable despite all its imperfections, deems important in the study of deterrence. In fact, at this point, this book is not introducing much new in this respect, but rather integrating dispersed concepts into a single research framework. The critical integration that allows addressing the problem of the analytical primacy of nuclear deterrence follows. Apart from cases where military factors did not play any role, which I do not consider as deterrence situations, the effects of a deterrer's military threats can be attributed to one or more of the following four categories: nuclear retaliation; *nuclear denial*; *conventional denial*; and conventional retaliation. In this step, the established division of deterrence between denial and punishment that was introduced by Glen Snyder already in the 1960s is utilized in a similar fashion as in the original, yet almost forgotten, article of Samuel Huntington.[71] This step overcomes the all-too-common bias that in fact reduces the study of nuclear deterrence to punishment and the study of conventional deterrence to denial.

Nuclear retaliation corresponds with the most traditional meaning of deterrence in public discourse. The threat is tailored in a way to punish the aggressor without trying to change his ability to grab the benefits of the aggression, at least in the short term. However, it signals that the benefits will not be worth the costs. The threat of nuclear retaliation that dominated Cold War thinking underpinned the logic of MAD, as well as the logics of existential deterrence and proliferation optimism.[72] Here the challenger's plans are altered because the deterrer threatens to respond with nuclear weapons against something that is valuable to the challenger, most usually against his cities. Thus, Waltz argues that Israel would not risk the destruction of Haifa and Tel Aviv, Libya would not risk Benghazi and Tripoli. He asks what kind of political objective would be worth risking "Vladivostok, Novosibirsk, and Tomsk, with no way of being sure that Moscow would not follow."[73]

The other type of nuclear deterrence is less developed, as denial strategies are traditionally more associated with conventional weapons. Glenn Snyder argued that denial is accomplished with the military forces that can block the challenger from accomplishing his goal, leaving him without the benefits of the aggression.[74] This text modifies Snyder substantially, as it draws the line between denial and punishment depending on the objective of the threats.[75] Such a step is not unreasonable. For Snyder, only a threat of successful denial of the enemy's objective makes denial. An unsuccessful attempt to block the objective will belong to punishment. But prior to the outbreak of hostilities, possible success cannot often be estimated by the challenger. Could NATO stop a Soviet invasion by a combination of conventional and tactical nuclear weapons? Neither NATO, nor the Soviets could know. Snyder's borderline is blurred. But the threat was there, and clearly it had a completely different logic then the threat of punishment of Soviet Union by nuclear strikes on Soviet cities. The former suffered from a much smaller credibility problem because it was clearly related to the attempt to deny the Soviet objective and not necessarily to escalate the conflict to a level that would be more harmful for both sides.

Conventional denial accounts for the most common perception of conventional deterrence. It corresponds with Snyder's image of conventional forces "which can block the enemy's military forces from making territorial gains."[76] Yet conventional denial is not necessarily related to territorial gains. In fact, in neither of my cases did the challenger primarily seek to alter the territorial status quo. This does not mean that denial could not work. Deterrence by denial utilizes the threat that the objective will not be achieved. Anti-aircraft batteries and SAM missiles around nuclear complexes are often not only intended to stop the approaching aircraft, but also to send the message that aircraft will be stopped, hence the challenger should drop his plans as futile. In search of deterrence, actors often declare their readiness to defend and in fact often exaggerate this readiness; they specify the means deployed even though by providing this information to the enemy the tactical surprise may be lost if deterrence fails.

Similarly to nuclear denial, conventional retaliation has mostly avoided scholarly attention. Usually it was assumed that only nuclear retaliation with long-range bombers or missiles can be achieved. This pitfall probably goes back to the way strategic nuclear weapons and ICBMs changed the traditional Second World War-like image of war. Against ICBMs, defense suddenly seemed impossible and retaliation of an unprecedented scale certain. On the other hand, conventional retaliation allegedly required a slow onslaught of the enemy on the battlefield and a traditional-like march into the enemy's territory which would end up in the capture of his capital. Naturally, only the conventionally stronger side could achieve that. Samuel Huntington challenged this view in his contribution to the debate about conventional deterrence that emerged briefly in the 1980s and is now, regrettably, almost forgotten. Huntington's suggestion for NATO was to threaten to respond to a Soviet attack on Western Europe by its own offensive into Soviet-controlled Eastern Europe. This step was supposed to eliminate the deficiency of conventional deterrence by imposing uncertainty on the enemy's plans, making him risk not only the failure of his plans, but also additional cost.[77]

Drawn from a number of theoretical accounts of international politics, the aforementioned concepts make a comparative framework of this study. Yet with the notorious secrecy that surrounds national security matters in general, and nuclear weapon related issues in particular, highly accurate data for all the segments of this framework may not be always available. Values can be accurately assigned with a high degree of confidence for some concepts, but only rough estimates will be available for others. Thus a simple qualitative system is employed to facilitate comparison.

This book mostly distinguishes between two or three possible values: none/limited/strong, or alternatively limited/strong or yes/no. The first type applies to levels of nuclear asymmetry, second-strike criterion, the challenger's technological advantage, third-party deterrence, alliance politics, nuclear taboo (where "emerging" stands for limited to keep in line with existing approaches), and opposition among influential decision makers. The second type covers the availability of information, centrality of conflict, perceived resolve of deterrer, institutionalization of mutual relations, history of hostility, last resort, perception of deterrer's rationality, international action legitimacy, domestic action legitimacy, nuclear retaliation, nuclear denial, conventional denial, and conventional retaliation. Admittedly this system sacrifices the bulk of uniqueness for easier cross-case comparison, yet the possible observation of general patterns of deterrence should provide a worthy compensation.

Table 1.1 Comparative framework

Concept	Description	Values
The deterrer's nuclear arsenal	Key physical attributes of respective nuclear complexes; weapon types; numbers; delivery vehicles; command, control, and communication systems	Actual number or a small interval estimate
Nuclear asymmetry	Situation when a substantially smaller (quantitatively) and less sophisticated nuclear force (qualitatively) faces a qualitatively and quantitatively larger nuclear force	None/limited/strong
Second-strike criterion	Nuclear posture that has the ability to survive enemy attack, make and communicate decision to retaliate, overcome enemy's active defense, and destroy a valuable target despite its passive defense	None/limited/strong
General conventional preponderance	Situation when one side's armed forces are in general substantially stronger in terms of numbers, technology, training, and employment strategies	Challenger/deterrer/rough balance
Theater conventional preponderance	Situation when one side's armed forces are substantially stronger in terms of numbers, technology, training, and employment strategies on a theater where targets valuable to the challenger can be found	Challenger/deterrer/rough balance
The challenger's technological advantage	Significant advantage of the challenger's major weapons systems that would be employed in case of conflict in terms of state-of-the-art sophistication over the deterrer's weapons systems	None/limited/strong
Availability of information	The challenger's knowledge about targets' location and defensive systems, which is established from sources that the challenger deems credible	Limited/strong
Centrality of conflict	Absolute importance of the conflict dyad between the challenger and the deterrer from the challenger's subjective perspective and its relative importance to other existing conflict dyads where the challenger is a party	Yes/no

continued

Table 1.1 Continued

Concept	Description	Values
Perceived resolve of the deterrer	The challenger's perception of the deterrer's commitment to fight should the deterrence fail	Limited/strong
Institutionalization of mutual relations	Degree of shared expectations about the requirements of stable deterrence and the existence of proven formal or informal communication channels	Limited/strong
History of hostility	Track record of conflict in the dyad that shapes the understanding of other side's intentions	Limited/strong
Last resort	Situation when the challenger sees only the options to strike, or to live up to the development he tries to prevent	Yes/no
Perception of the deterrer's rationality	The challenger's perception of the deterrer's rationality, particularly whether the challenger believes it possible to live with the nuclear-armed deterrer in the long term	Yes/no
Third-party deterrence	Threat of military involvement into the original conflict by a third party, most likely the allies of the original deterrer, that decisively influenced the challenger's decision	None/limited/strong
Alliance politics	Sensitivity of the challenger to possible impact of his action on the relations with his allies	None/limited/strong
International action legitimacy	The challenger's sensitivity to the international normative expectation of non-intervention	Limited/strong
Nuclear taboo	Normative prohibition on the use of nuclear weapons	None/emerging/strong

Domestic action legitimacy	Level of support for military solution among the challenger's population	Limited/strong
Opposition among influential decision makers	Negative view on military solution by influential part of the challenger's government	None/limited/strong
Regime type	The challenger's position on the democracy-nondemocracy axis and on the militarized-nonmilitarized axis	Democracy/autocracy and militarized/non-militarized
Nuclear retaliation	Threat that the deterrer will use its nuclear weapons against targets valuable to the challenger, except of targets that are directly related to pursuit of the challenger's objectives	Limited/strong
Nuclear denial	Threat that the deterrer will use its nuclear weapons against targets that are directly related to pursuit of the challenger's objectives in order to prevent him from attaining the objectives, or in order to make it unacceptably costly	Limited/strong
Conventional denial	Threat that the deterrer will use its conventional weapons against targets that are directly related to pursuit of the challenger's objectives in order to prevent him from attaining the objectives, or in order to make it unacceptably costly	Limited/strong
Conventional retaliation	Threat that the deterrer will use its conventional weapons against targets valuable to the challenger, except of targets that are directly related to pursuit of the challenger's objectives	Limited/strong

Selecting cases

This book employs the method of a comparative historical case study. Thus it is of little surprise that case selection is vital for its success. It must correspond with the scope of this research and allow control to be as careful as possible. Interest lies in the interplay of nuclear, conventional, and non-military factors in a deterrence situation between a powerful challenger and a deterrer with a small nuclear arsenal. Thus, in the first place, the book needs cases from dyads between big and small. Usually this will be an emerging nuclear state and an established power. The actual status of an established power is easy to recognize. Recognizing the status of a small nuclear state is more demanding. I argue that the right description is one where at least some influential decision makers in the challenging state believe that the deterrer possesses at least one nuclear weapon. The actual existence of the weapon is less important. Advanced nuclear programs make it almost impossible for opponents to determine when the nuclear threshold has been crossed exactly. In a later phase of nuclear weapons development, "whether or not it [country] has nuclear weapons may not be surely known."[78] Nevertheless, if the challenger believes in the weapon's existence, he will behave as if it was real.

I also argue that the mechanisms of deterrence should be best visible in acute crises. This corresponds with the traditional division of deterrence into two types, immediate and general. Patrick Morgan, the author of this useful distinction, describes immediate deterrence as "the relationship between opposing states where at least one side is seriously considering an attack while the other is mounting a threat of retaliation in order to prevent it," while general deterrence "relates to opponents who maintain armed forces to regulate their relationship even though neither is anywhere near mounting an attack."[79] Both types require a military component, yet in a general deterrence situation, the mechanisms will be often unobservable and hence difficult to study. In general deterrence, hostility makes conflict possible. In immediate deterrence, leaders review plans, make considerations and deliberations, militaries take preparations, threats are issued and these steps can be observed and traced in empirical evidence. Therefore, I limit my attention to immediate deterrence cases.

But immediate deterrence situations are not common in the real world. General deterrence is more common and only rarely degenerates into immediate deterrence. When an immediate deterrence situation arises, it usually reflects a failure of general deterrence, not just some immediate highly conflictual developments. To meet the criteria of immediate deterrence a situation requires hostility between the two actors where one side seriously considers attack and the other, aware of the threat, must issue or at least tacitly imply the counter-threat of military reaction. The challenger's leaders must then drop their plans primarily because of the deterrer's threats.[80] In a sense, only three of Morgan's conditions – the challenger considering an attack, the deterrer knowing about the threat, and the deterrer communicating a counter-threat – are necessary for an immediate deterrence situation. The last condition – abandonment of the attack primarily due to the deterrer's threat – should be broadened to include deterrence failures, the moments when the challenger does not drop his plan but proceeds to attack. In fact those were the situations which most clearly required the deterrer's military threat to prevent the outbreak of hostilities.

The aforementioned conditions fairly limit the number of cases that fit my research design. Also the most recent cases are unlikely to be data-rich as most information remains classified. This makes large-N statistical research unrealistic but fits well into my comparative case study plan. Due to the limited number of cases, my study is able to cover almost the entire population of cases from the early nuclear era till the 1990s. Thus, the following cases are examined:

- the United States and China in the early 1960s,
- the Soviet Union and China in the late 1960s,
- Israel and Iraq in 1977–1981 (Osiraq),
- the United States and North Korea in 1994,
- the United States and the Soviet Union in 1962 (the Cuban Missile Crisis, the control case).

In terms of general deterrence, all the cases included in the study are in fact very much mutual deterrence situations. In all of them both sides saw the other side as hostile and aggressive. In a sense, both can be seen as a challenger and deterrer at the same time. But as I limit myself to immediate deterrence situations I can analytically label one side – the one that was seriously considering an attack – the challenger and the other the deterrer. At least hypothetically, the other side could have been also considering an attack at the very same time. Empirically the cases are apparently complex. But in these complex cases I am only interested in operations of immediate deterrence in the minds of those who were being deterred by small or emerging nuclear arsenal. All the cases unquestionably meet the three critical criteria for immediate deterrence.

They also allow for a fairly reasonable variety across time, space, type of actors, and other factors to enhance research validity. For instance, the challengers are represented by both democratic and non-democratic states. Cases from the early nuclear age and from the post-Cold War RMA age are included to control for the effect of changing technology, one of the most important enablers of successful prevention. Unequivocal regional powers such as China, global superpowers such as the Soviet Union, and far less influential players such as Iraq and North Korea can be found in the list of deterrers. The Cuban Missile Crisis, while also marked with a substantial degree of asymmetry between the two sides' nuclear arsenals, offers at least limited control in terms of quality of asymmetry. While the disparity existed between the two superpowers in nuclear realm, it was qualitatively different to other cases particularly on the sub-strategic level. Deterrence successes are listed, which is no surprise as it was already mentioned above that no unquestionable, clear-cut example of nuclear deterrence failure exists. Nonetheless, the failure of deterrence between Iraq and Israel in 1981 offers an at least imperfect example of deterrence failure and is included to allow for the strongest possible control.

Of course, this research strategy and case selection has its tradeoffs. This study cannot answer certain important questions nor does it try to. It cannot say much about general deterrence, nor even about the way how a general deterrence situation deteriorates into immediate deterrence. Being able to do this would be extremely beneficial for scholarship as it is the other less researched part of deterrence. But it is beyond the scope of this book. Second, this study cannot claim high validity for small-to-small dyads. The fact that a small nuclear arsenal cannot unconditionally deter a superpower that employs a large, superbly trained and technologically sophisticated military force does not mean that a small regional adversary cannot be deterred by the same arsenal. This would be a question for further research.

Despite the aforementioned limitations, this study still bears an extremely important theoretical insight and equally valuable policy recommendations. It must be admitted that the number of cases remains rather low, some controls are rather weak, and generalizations cannot be made with unequivocal certainty. Yet this is still by far the best that can be done with the available data.

Notes

1 Bernard Brodie, 'Strategy as a Science', *World Politics*, 1/4 (1949) 467–488.
2 Richard K. Betts, "Should Strategic Studies Survive?," *World Politics*, 50/1 (1997) 13; Albert Wohlstetter and Fred S. Hoffman, "Defending a Strategic Force After 1960," *D-2270* (Santa Monica CA: RAND, 1954).
3 Albert Wohlstetter, "Delicate Balance of Terror," *Foreign Affairs*, 37/2 (1959) 211–234.
4 Glenn Snyder, *Deterrence and Defense* (Princeton: Princeton University Press, 1961).
5 Thomas C. Schelling, *Arms and Influence* (New Haven: Yale University Press, 1966).
6 Other notable examples are Bernard Brodie, *The Absolute Weapon* (New York: Harcourt, 1946); Bernard Brodie, *Strategy in the Missile Age* (Princeton: Princeton University Press, 1959); Herman Kahn, *On Thermonuclear War* (Princeton: Princeton University Press, 1960); Thomas C. Schelling, *The Strategy of Conflict* (Cambridge: Harvard University Press, 1960).

7 Christopher H. Achen and Duncan Snidal, "Rational Deterrence and Comparative Case Studies," *World Politics*, 41/2 (January 1989) 143; notable examples include Frank C. Zagare, *The Dynamics of Deterrence* (Chicago: University of Chicago Press, 1987); Frank C. Zagare and D. Marc Kilgour, *Perfect Deterrence* (Cambridge: Cambridge University Press, 2000); Robert Powell, "The Theoretical Foundations of Strategic Nuclear Deterrence," *Political Science Quarterly*, 100/1 (Spring 1985) 75–96; Robert Powell, *Nuclear Deterrence Theory: The Search for Credibility* (Cambridge: Cambridge University Press, 1990).

8 Achen and Snidal, "Rational Deterrence and Comparative Case Studies," 505.

9 Alexander L. George and Richard Smoke, Deterrence in American Foreign Policy (New York: Columbia University Press, 1970); Robert Jervis, *Perception and Misperception in International Politics* (Princeton: Princeton University Press, 1976; Richard Ned Lebow, *Between Peace and War: The Nature of International Crisis* (Baltimore: The Johns Hopkins University Press, 1981); Robert Jervis, Richard Ned Lebow and Janice Gross Stein (eds), *Psychology and Deterrence* (Baltimore: The Johns Hopkins University Press, 1985).

10 A good overview can be found in The Rational Deterrence Debate: A Symposium published in the *World Politics* 4/2 (January 1989) with contributions by Christopher H. Achen and Duncan Snidal, "Rational Deterrence Theory and Comparative Case Studies," 143–169; Alexander L. George and Richard Smoke, "Deterrence and Foreign Policy," 170–182; Robert Jervis, "Rational Deterrence: Theory and Evidence," 183–207; Richard Ned Lebow and Janice Gross Stein, "Rational Deterrence Theory: I Think, Therefore I Deter," 208–224; George W. Downs, "The Rational Deterrence Debate," 225–237.

11 E.g., Pierre Galloise, *Stratégie de l'âge nucléaire* (Paris: Calmann-Lévy, 1960); André Beaufre, *Deterrence and Strategy* (New York: Frederick A. Praeger, 1966).

12 Albert Wohlstetter, "Nuclear Sharing: NATO and the N+1 Country," *Foreign Affairs*, 39/3 (April 1961) 355–387.

13 McGeorge Bundy, "Existential Deterrence and its Consequences" in Douglas MacLean (eds), *The Security Gamble: Deterrence Dilemmas in the Nuclear Age* (Totowa: Rowman and Littlefield, 1984) 3–13.

14 Vipin Narang, "What Does It Take to Deter? Regional Power Nuclear Posture and International Conflict," *Journal of Conflict Resolution*, 57/3 (June 2013) 479.

15 Kenneth N. Waltz, "The Spread of Nuclear Weapons: More May Better," *Adelphi Papers*, Number 171 (London: International Institute for Strategic Studies, 1981).

16 John J. Mearsheimer, "Back to the Future: Instability in Europe After the Cold War," *International Security*, 15/1 (Summer 1990) 38; Merasheimer, "The Case for a Ukrainian Nuclear Deterrent"; Stephen Van Evera, "Primed for Peace: Europe after the Cold War," *International Security*, 15/3 (Winter 1990–1991) 12–14; Peter R. Lavoy, "The Strategic Consequences of Nuclear Proliferation," *Security Studies*, 4/4 (Summer 1995) 695–753; David J. Karl, "Proliferation Pessimism and Emerging Nuclear Powers," *International Security*, 21/3 (Winter 1996–1997) 87–119; Victor D. Cha, "The Second Nuclear Age: Proliferation Pessimism Versus Sober Optimism in South Asia and East Asia," *Journal of Strategic Studies*, 24/4 (2001) 79–120.

17 Scott D. Sagan, "More Will Be Worse" in Scott D. Sagan and Kenneth N. Waltz, *The Spread of Nuclear Weapons: A Debate Renewed* (New York: W.W. Norton, 2003) 46–87; Peter D. Feaver, "Proliferation Optimism and Theories of Nuclear Operations," *Security Studies*, 2/3–4 (1993) 15–191; Peter D. Feaver, "Optimists,

Pessimists, and Theories of Nuclear Proliferation Management: Debate," *Security Studies*, 4/4 (Summer 1995) 754–772; Lyle J. Goldstein, *Preventive Attack and Weapons of Mass Destruction: A Comparative Historical Analysis* (Stanford: Stanford University Press, 2006).

18   Sagan, "More Will Be Worse," 53–63.
19   Sagan, "More Will Be Worse," 73–82.
20   Sagan, "More Will Be Worse," 63–72.
21   Feaver, "Proliferation Optimism and Theories of Nuclear Operations."
22   Scott D. Sagan, "Sagan Responds to Waltz," in Scott D. Sagan and Kenneth N. Waltz, *The Spread of Nuclear Weapons: A Debate Renewed* (New York: W.W. Norton, 2003) 159–166.
23   Jeffrey W. Knopf, "Recasting the Proliferation Optimism-Pessimism Debate," *Security Studies*, 12/1, 41–96; Matthew Kroenig, "Beyond Optimism and Pessimism: The Deferential Effects of Nuclear Proliferation," *Managing the Atom Working Paper No. 2009–14* (Cambridge: Harvard University, November 2009); also to some extent Feaver, "Proliferation Optimism and Theories of Nuclear Operations."
24   Sarah E. Kreps and Matthew Fuhrmann, "Attacking Atom: Does Bombing Nuclear Facilities Affect Proliferation?," *Journal of Strategic Studies*, 34/2 (2011) 161–187.
25   The former includes original contributions of both Waltz and Sagan and also early post-Cold War scholarship; the latter, for instance, Kenneth N. Waltz and Scott D. Sagan, "Indian and Pakistani Nuclear Weapons: For Better or Worse?" in Scott D. Sagan and Kenneth N. Waltz, *The Spread of Nuclear Weapons: A Debate Renewed* (New York: W.W. Norton, 2003) 88–124; Karl, "Proliferation Pessimism and Emerging Nuclear Powers"; Cha, "The Second Nuclear Age: Proliferation Pessimism Versus Sober Optimism in South Asia and East Asia."
26   Other authors have examined cases of preventive strikes and "near-misses" but they usually do not frame their exploration within the context of the debate between proliferation pessimists and optimists. See, e.g., Robert S. Litwak, "The New Calculus of Pre-emption," *Survival*, 44/4 (Winter 2002–2003) 53–80; Kreps and Fuhrmann, "Attacking Atom: Does Bombing Nuclear Facilities Affect Proliferation?"; Thérèse Delpech, *Nuclear Deterrence in the 21st Century: Lessons from Cold War for a New Era of Strategic Piracy* (Santa Monica: RAND Corporation, 2012).
27   Goldstein, *Preventive Attack and Weapons of Mass Destruction*, 151, 152.
28   Richard K. Betts, "Conventional Deterrence: Predictive Uncertainty and Policy Confidence," *World Politics*, 37/2 (January 1985) 155.
29   John J. Mearsheimer, *Conventional Deterrence* (Ithaca: Cornell University Press,1983) 23–24.
30   Lawrence Freedman, *Strategy: A History* (Oxford: Oxford University Press, 2013) 194.
31   The most systematic treatment is offered by Mearsheimer's *Conventional Deterrence*. For other important contributions see, e.g., Betts, "Conventional Deterrence"; Samuel P. Huntington, "Conventional Deterrence and Conventional Retaliation in Europe," *International Security* 8/3 (Winter 1983–1984) 32–56.
32   Patrick M. Morgan, *Deterrence: A Conceptual Analysis* (London: Sage, 1977) 39.
33   Alexander L. George and Andrew Bennet, *Case Studies and Theory Development in the Social Sciences* (Cambridge: MIT Press, 2005) 220.
34   See Nik Hynek and Michal Smetana (eds), *Global Nuclear Disarmament* (London: Routledge, 2016).

35　Tim Dunne, Lene Hansen, and Colin Wight, "The End of International Relations Theory?," *European Journal of International Relations*, 19/3 (2013) 405–425; Christian Reus-Smit, "Beyond metatheory?" *European Journal of International Relations*, 19/3 (2013) 589–608.

36　Scott Sagan, "Why Do States Build Nuclear Weapons," *International Security*, 21/3 (Winter 1996–1997) 54–86.

37　Bundy, "Existential Deterrence and its Consequences."

38　Tom Sauer, "A Second Nuclear Revolution: From Nuclear Primacy to Post-Existential Deterrence," *Journal of Strategic Studies*, 32/5 (October 2009) 749.

39　Henry Sokolski (ed.) *Getting MAD: Nuclear Mutual Assured Destruction, Its Origins and Practice* (Carlisle: Strategic Studies Institute, U.S. Army War College, 2004).

40　Colin S. Gray, "Nuclear Strategy: The Case for Theory of Victory," *International Security*, 4/1 (1979) 54–87; Collin S. Gray and Keith Payne, "Victory is Possible," *Foreign Policy*, 39 (1980) 14–27.

41　In fact, the second strike is the only optimists' requirement for deterrence not to fail due to a conscious decision by the challenger. Two other requirements include the ability to avoid launches on false alarm and the ability to avoid unauthorized use. These two are directly related to the most serious concerns of organizational theory-based pessimism. Yet the fear of unauthorized and accidental use has little to do with deterrence; it is related to other possible negative consequences of proliferation. Waltz, "More May Be Better," 20.

42　Waltz, "More May Be Better," 19.

43　Waltz, "More May Be Better," 20.

44　Wohlstetter, "Delicate Balance of Terror," 216.

45　See Mearsheimer, *Conventional Deterrence*, 23–66; Stephen Biddle, *Military Power: Explaining Victory and Defeat in Modern Battle* (Princeton: Princeton University Press, 2006) 14–51; Daryl G. Press, "Lessons from Ground Combat in the Gulf: The Impact of Training and Technology," *International Security*, 22/2 (Fall 1997) 137–146; Daryl G. Press, "The Myth of Air Power in Persian Gulf War and the Future of Warfare," *International Security*, 26/2 (Fall 2001) 5–44.

46　See Jan Ludvik, "The Poverty of Statistics: Military Power and Strategic Balance," *Central European Journal of International and Security Studies*, 8/4 (2014) 152–166.

47　See Thomas A. Keaney and Eliot A. Cohen, *Revolution in warfare? Air Power in the Persian Gulf* (Annapolis: Naval Institute Press, 1995) Chapters 9, 10; Martin van Creveld, *Technology and War: From 2000 B.C. to the Present* (New York: The Free Press: 1991).

48　Michel Fortman and Stefanie von Hlatky, "Revolution in Military Affairs: Impact of Emerging Technologies on Deterrence," in T.V. Paul, Patrick Morgan and James J. Wirtz (eds), *Complex Deterrence: Strategy in Global Age* (Chicago: University of Chicago Press, 2009) 304–319.

49　Charles L. Glaser and Steve Fetter, "Counterforce Revisited: Assessing the Nuclear Posture Review's New Missions," *International Security*, 30/2 (Fall 2005) 89–95.

50　See Daniel R. Lake, "The Limits of Coercive Airpower: NATO's 'Victory' in Kosovo Revisited," *International* Security, 34/1 (Summer 2009) 96.

51　Waltz, "More May Be Better," 19–20.

52　Klaus Knorr and Patrick Morgan (eds), *Strategic Military Surprise: Incentives and Opportunities* (New Brunswick: Transaction Books, 1983).

53　Powell, *Nuclear Deterrence Theory*, Chapter 2.

54 Powell, "Nuclear Deterrence Theory, Nuclear Proliferation, and National Missile Defense," 89.
55 Schelling, *Strategy of Conflict*, 187.
56 Robert Jervis, "Introduction: Approach and Assumptions," in Robert Jervis, Richard Ned Lebow and Janice Gross Stein (eds), *Psychology and Deterrence* (Baltimore: The Johns Hopkins University Press, 1985) 1–12.
57 Robert Jervis, "Perceiving and Coping with Threat," in Richard Ned Lebow and Janice Gross Stein (eds), *Psychology and Deterrence* (Baltimore: The Johns Hopkins University Press, 1985) 14.
58 For misperception and deterrence see Jervis, *Perception and Misperception in International Politics* particularly Chapter 3; for a broader overview of political psychology's contributions to IR, see Rose McDermott, *Political Psychology in International Relations* (Ann Arbor: University of Michigan Press, 2009).
59 Darryl Howlett, "New Concepts of Deterrence," in William C. Potter and John Simpson (eds), *International Perspectives on Missile Proliferation and Defenses, Occasion Paper No. 5* (Monterey: Center for Nonproliferation Studies, 2001) 20–21.
60 White House, *National Security Strategy of the United States of America* (Washington D.C.: White House, 2002) 13–16.
61 See Derek D. Smith, *Deterring America* (West Nyack: Cambridge University Press, 2002) 3–12.
62 See Paul K. Huth, *Extended Deterrence and the Prevention of War* (New Haven: Yale University Press, 1988); Alastair Smith, "Extended Deterrence and Alliance Formation," *International Interactions*, 24/4 (1998) 315–343.
63 Timothy W. Crawford, *Pivotal Deterrence: Third-Party Statecraft and the Pursuit of Peace* (Ithaca: Cornell University Press, 2003).
64 Ian Hurd, "Legitimacy and Authority in International Politics," *International Organization*, 53/2 (1999) 379.
65 Thomas M. Nichols, *Eve of Destruction: The Coming Age of Preventive War* (Philadelphia: University of Pennsylvania Press, 2008) 14–39.
66 Nichols, *Eve of Destruction*.
67 Nina Tannewald, "The Nuclear Taboo: The United States and the Normative Basis of Nuclear Non-Use," *International Organization*, 53/3 (Summer 1999) 434. For more skeptical views see T.V. Paul, "Taboo or Tradition? The Non-Use of Nuclear Weapons in World Politics," *Review of International Studies*, 36/4 (2010) 853–863; Daryl G. Press, Scott D. Sagan, and Benjamin A. Valentino, "Atomic Aversion: Experimental Evidence on Taboos, Traditions, and the Non-Use of Nuclear Weapons," *American Political Science Review*, 107/1 (February 2013) 188–206.
68 Valerie M. Hudson, "Foreign Policy Analysis: Actor-Specific Theory and the Ground of International Relations," *Foreign Policy Analysis*, 1/1 (2005) 1–30; also see Jack S. Levy, "Domestic Politics and War," *Journal of Interdisciplinary History*, 43/4 (Spring 1988) 653–673.
69 See Michael Doyle, "Liberalism and World Politics," *American Political Science Review*, 80/4 (December 1986) 1151–1169; Bruce Russett, *Grasping the Democratic Peace Theory* (Princeton: Princeton University Press, 1993); Karen A. Rasler and William R. Thompson, *Puzzles of the Democratic Peace: Theory, Geopolitics and the Transformation of World Politics* (Gordonsville: Palgrave Macmillan, 2005); for another notable example of domestic regime's influence on the likelihood of war and peace, see Edward D. Mansfield and Jack Snyder, "Democratization and the Danger of War," *International Security*, 20/1 (Summer 1995) 5–38 on democratizing nations' increased taste for conflict.

70 Christopher Layne, "Kant or Cant: The Myth of Democratic Peace," *International Security*, 19/2 (Autumn 1994) 5–49.

71 Snyder, *Deterrence and Defense*; Huntington, "Conventional Deterrence and Conventional Retaliation in Europe."

72 See Lawrence Freedman, *The Evolution of Nuclear Strategy* (Basingstoke: Palgrave Macmillan, 2003); Budny, "Existential Deterrence and its Consequences."

73 Waltz, "More May Be Better," 22–23.

74 Glenn H. Snyder, "Deterrence and Power," *The Journal of Conflict Resolution*, 4/2 (June 1960) 165.

75 A similar step is taken by Robert F. Trager and Dessislava P. Zagorcheva, "Deterring Terrorism: It Can Be Done," *International Security*, 30/3 (Winter 2005/2006) 87–123, who argue that by increasing the chance that a terrorist attack will not succeed the terrorists can be deterred by denial even though the protective measures can never complete. The argument is the same as mine. The threat that the terrorist will be stopped creates deterrence by denial. There is no need to be sure that they will be stopped. This is not to say that deterrence success is guaranteed. The threat may not be robust enough to deter. But it clearly has a completely different logic than the threat of punishment.

76 Snyder, "Deterrence and Power," 165.

77 Huntington, "Conventional Deterrence and Conventional Retaliation in Europe," 37.

78 Kenneth N. Waltz, "More May Be Better," in Scott D. Sagan and Kenneth N. Waltz, *The Spread of Nuclear Weapons: A Debate Renewed* (New York: W.W. Norton, 2003) 18.

79 Patrick M. Morgan, *Deterrence: A Conceptual Analysis* (Sage: London 1977) 28.

80 Morgan, *Deterrence*, 31–40.

2   The United States and China, 1959–1966

As with most nuclear weapons-related matters, a lot remains unclear even in this book's historically oldest case study. The United States declassified voluminous and valuable governmental sources, yet even with them only an incomplete picture can be reconstructed. Some parts are missing, erased from the declassified materials, or remain classified. Important Chinese and Soviet sources are even less available. However, despite some blank spots remaining on the timeline, the available insight into the challenger's thinking and decision making is impressive and illustrative. Now, there is a lot that we know.

The plot

When exactly the People's Republic made the critical decision remains unclear. Most authors put the date at January 1955, when Mao Zedong discussed the matter with several leading scientists.[1] Yet initial probes were probably made earlier.[2] It is clear that by the middle of 1955 Mao had made the decision that it was necessary for his country to go nuclear. By that time China had fairly decent experience of living with the formidable threat of a hostile nuclear power. Both Truman's and Eisenhower's administrations tried to use ambiguous nuclear threats to get China's compliance during the Korean War.[3] Another lesson followed just a year after the end of war in Korea. A crisis between the United States and China over the Chinese nationalist-controlled offshore island occurred in the Taiwan Strait. In the heat of this crisis, President Eisenhower and his Secretary of State Dulles signaled to the PRC that should the Americans end up compelled to defend Chiang Kai-shek's positions the atomic bombs would be part of the battle plan. While the Chinese publicly pronounced their disrespect of nuclear "paper tigers," and also while it does not appear that their behavior was greatly influenced by nuclear threats per se, they felt the need to build their own nuclear option, at least for political reasons.

China started to materialize its decision to get nuclear weapons in July 1955 when the CCP's Politburo appointed a small group of three of its members, Chen Yun, Marshal Nie Rongzhen, and Bo Yibo, to take care of nuclear research and development. The group subsequently established the Second Ministry of Machine Building, in charge of atomic bomb development, and the Fifth Academy, responsible for missile and space technologies. With a variety of Soviet help and advice, the construction of key facilities, including the Baotou Nuclear Fuel Component Plant, which would produce uranium tetra fluoride, and the Lanzhou Gaseous Diffusion Plant, which would enrich uranium, began in 1958. Two years later, in 1960, construction started at the Jiuquan Atomic Energy Complex, with the first plutonium production reactor, and at the test site in Lop Nur.[4] The location of most components in remote northern and western parts of China was hardly a coincidence. It was selected with the American threat in mind and, for further protection, the program ran under the highest possible degree of secrecy.[5]

At least initially, the Chinese hoped for an easy path to the bomb with substantial assistance of their Soviet allies. Already in February 1953, a Chinese scientific delegation arrived in Moscow to secure Stalin's support. Yet the timing was not on the Chinese side. The Soviet leader's health was failing. He died on March 5 before any deal could be closed and the issue was left to his heirs to be resolved. His successor, Nikita Khrushchev, when approached with a request to help with Chinese nuclear research and development in September 1954, immediately refused.[6]

Khrushchev was probably never enthusiastic about granting nuclear assistance to Beijing yet he moderated his earlier position and, for political reasons, the PRC's nuclear program received substantial technical assistance from Moscow from 1955 to 1959.[7] According to some accounts, the deal even included the delivery of a "sample atomic bomb" and allegedly the transport of this bomb was prepared, with a special railroad carriage readied to leave and stopped only by a last-moment decision of the Politburo.[8] Soviet leaders were not ready to go that far with the troublesome ally whom they did not really trust, at a time when a rift between the two countries was apparently opening. By June 1959, all of the Soviet advisors left the Chinese nuclear program; however, what was left was already a fairly sophisticated program.

Despite Chinese concealment, the United States learned fairly soon about the existence of the program. The possible reality of a Chinese nuclear program was already mentioned in a National Intelligence Estimate (NIE) from July 1959 and an aerial photograph of Lanzhou was taken in September.[9] The immediate reaction of Eisenhower's administration appeared calm, even though it is possible that the evidence initially escaped the attention of top decision makers. Secretary of State Herter – probably uninformed – raised the issue that the PRC had possibly acquired nuclear capability during a National Security Council meeting on June 22, 1960 mentioning that he

noticed a report of this possibility in a British newspaper. Yet CIA representatives could only confirm that the program existed and that available evidence was too scarce to make informed estimates.[10] The CIA apparently
followed the development of the Chinese nuclear program and covered the
issue in NIEs,[11] but no conclusive evidence could be given to the Eisenhower
administration to discuss what to do with the Chinese nuclear threat. Arguably the problem was left to the incoming administration.

It appears likely that the new president soon learned about the Chinese
ambitions and, according to Walt W. Rostow, he was chilled by the fact that
"the biggest event of the 1960s [might] well be the Chinese explosion of a
nuclear weapon."[12] Kennedy clearly felt the need to do something to prevent
the Chinese communists from being nuclear-armed. Thus, during a June
1961 summit with Khrushchev in Vienna, Kennedy probed the Soviet leader's position. Despite the rift between the two communist powers apparently
widening, Khrushchev, afraid of being perceived as "soft" on imperialism,
was not ready to join Kennedy's anti-nuclear-China camp.[13] However,
Kennedy was not discouraged by the Soviet rejection.

In January 1962, Kennedy tasked the NSC to deal with a "special unre-
solved problem" of Chinese nuclear weapons.[14] After a mid-1962 revision of
western policy on nuclear testing, the administration came to understand that
a test ban treaty with the Soviets should be a key diplomatic step in mitigating China's nuclear ambitions. On July 30, 1962, Assistant Secretary of
Defense for International Security Affairs Paul Nitze informed the President
and his top advisers that a "test ban would be a necessary, but not a sufficient, condition for inhibiting this proliferation, and that to prevent it would
require collaboration by the U.S. and USSR."[15] Yet the road to a test ban was
difficult. The French refusal to join made it difficult for the Americans to
convince the Soviets to join, and it was clear that Beijing would not feel the
need to adhere to any such treaty. Furthermore, the Soviets were reluctant to
allow onsite inspections in the USSR, thus the two sides could not find a
satisfactory form of verification. The differences turned out to be insuperable
obstacles to a complete test ban. In July 1963, when W. Averell Harriman, at
the time Under Secretary of State for Political Affairs, arrived in Moscow to
negotiate the treaty, he soon discovered that Khrushchev could not be moved
on the inspection issue. Yet, after the Cuban Missile Crisis, both Kennedy
and Khrushchev felt the need to act in the realm of arms control. Therefore,
the parties in Moscow quickly moved the negotiations to an alternative
partial option, concluding the "Treaty Banning Nuclear Weapon Tests in the
Atmosphere, in Outer Space and Under Water" in less than two weeks.[16]

To Kennedy, the negotiations in Moscow probably had another side. They
offered an opportunity to probe the Soviet position on the Chinese nuclear
program again. Kennedy informed Harriman about his view of the Chinese
nuclear threat and gave him the following instructions:

I remain convinced that Chinese problem is more serious than Khrushchev's comments in first meeting suggest, and believe you should press question in private meeting with him. I agree that large stockpiles are characteristic of U.S. and USSR only, but consider that relatively small forces in hands of people like CHICOMS [Chinese Communists] could be very dangerous to us all. Further believe even limited test ban can and should be means to limit diffusion. You should try to elicit Khrushchev's view of means of limiting or preventing Chinese nuclear development and his willingness either to take Soviet action or to accept U.S. action aimed in this direction.[17]

According to Chang and also Burr and Richelson, the President's instruction to Harriman aimed at discussing the possibility of a joint political and, if necessary, military action. On the contrary, Freedman argues that the President himself had no specific idea in mind and was not really pressing for a military strike.[18] What Kennedy really meant when putting the word "action" on paper is unclear.[19]

While both interpretations are plausible, putting a military option on the table appears more likely. In January 1963, National Security Advisor McGeorge Bundy informed the Director of Central Intelligence John McCone that Cuba and the Chinese nuclear program were the "two issues foremost in the minds of the highest authority and therefore should be treated accordingly by CIA."[20] By that time, the CIA had significantly extended its coverage of Chinese nuclear facilities by U.S. U-2s piloted by Taiwanese pilots and CORONA satellites, providing improved information including that necessary for military targeting. In February, Nitze requested the JCS study the options of how to persuade, pressure, or coerce the Chinese into signing the test ban treaty assuming the Soviet Union would join the U.S. action, or at least not interfere.[21] Even though Nitze belongs to the few formative figures of the Cold War whose influence often exceeded their formal position, it is unlikely that he could have made such a request without some approval from his superiors.

In response, the Joint Chiefs prepared a top-secret report on possible options to take out the Chinese nuclear program, exploring various actions, including diplomatic and military. Considering the military scenarios, the chiefs concluded that the United States: "has the capability of destroying either by conventional or nuclear air attack the identified CHICOM atomic energy facilities."[22] Yet it was also expected that both conventional and a limited nuclear strike would invite retaliation and escalation ranging from propaganda to direct military aggression against Asian countries allied to the West. The United States was believed to be able to counter any such military action initiated by the PRC, but probably they would have to resort to the use of tactical nuclear weapons. Furthermore, the chiefs questioned Nitze's assumption about Soviet behavior and estimated that the Soviet Union would intervene on behalf of communist China in the case of more aggressive American actions. Soviet consent with the action, therefore, seemed to be a precondition for success. But as Harriman learned in Moscow, the Soviets were not interested in discussing anti-Chinese action.

Khrushchev may not have been interested in discussing anti-Chinese action, yet it does not seem that he went in the opposite direction and threatened a hostile Soviet reaction to American action. Certainly Washington did not feel that way. Harriman informed Washington about his conversation with Khrushchev on July 27.[23] A few days later, on July 31, Acting Assistant Secretary of Defense for International Security William Bundy requested a "contingency plan for an attack with conventional weapons on Chinese Communist nuclear weapons production facilities designed to cause severest impact on and delay in the Chinese program."[24] On December 14, the JCS responded that conventional action was feasible, but they recommended considering tactical nuclear weapons for such an attack.[25] The Chiefs also prepared the actual contingency plan that was marked Plan 94 and which provided "for overt operations employing U.S. forces in air strikes against a ChiCom nuclear production facility."[26]

Military options to hamper the PRC's nuclear program clearly received substantial attention during Kennedy's last year in office. In September, not long after Harriman's mission to Moscow, an opportunity arose when Chiang Kai-shek's son, General Chiang Ching-kuo, arrived in Washington. Chiang first discussed the matters of PRC's nuclear program with McGeorge Bundy and a day later directly with the President. The nationalist leader appealed to Americans to support the GRC's (Government of Republic of China) policies arguing that the PRC was at its weakest point and was not going to receive military help from the Soviets. Furthermore he claimed that the GRC "has located missile sites and atomic installations on the China mainland and desires to work with the United States on ways and means to remove these and restrain their expansion" and offered that the GRC would "assume full political responsibility for this action, expecting only transportation and technical assistance from the United States."[27] Thus during Chiang's discussion with Kennedy, the President raised the question if it "would be possible to send 300 to 500 men by air to such distant Chinese Communist atomic installation as that at [Baotou], and whether it was not likely that the planes involved would be shot down."[28] At that point Chiang agreed that such an operation had been discussed the day before and that it was feasible.[29] Kennedy probably had his doubts about it, referring to the Bay of Pigs fiasco in the flow of conversation with Chiang, yet he did not prevent the establishment of a planning group to study the feasibility of a Nationalist attack on the PRC's nuclear installation during a meeting between Chiang and DCI McCone in the next days.[30]

In the meantime, the opposition to a direct attack arose within the government. The State Department's Policy Planning Staff (PPS) thoroughly reviewed the likely impact of the Chinese possession of nuclear weapons. Several studies were prepared by its East Asian specialist Robert Johnson and received a favorable impression from his superiors, including PPS director Walt W. Rostow and Secretary of State Dean Rusk. Johnson argued that a nuclear-armed China would be rather more cautious, the first use of nuclear weapons by China would be unlikely except for a regime threatening to attack the mainland, which the Americans were not planning, and no restraints would be put on American ability to oppose communist aggression in Asia. Thus Johnson suggested that a proper American reaction should be political action and greater readiness to challenge possible Chinese conventional actions with U.S. conventional force.[31] The first 200-page-long study was ready by July 1963 and Rostow, probably influenced by its reading, informed Harriman before his mission to Moscow that the Chinese nuclear capacity did not appear particularly threatening.[32] In October, Johnson, working in cooperation with others across the civil service, prepared a shorter version for wider circulation, yet it appears unlikely that Kennedy was acquainted with Johnson's study before his death in November.

Kennedy's death makes it hard to distinguish whether the subsequent cooling down in the plans to destroy the emerging Chinese nuclear arsenals should be ascribed to the reassessment of the Chinese threat pressed by the State Department, or to the changed personality of the president. Certainly President Johnson abandoned Kennedy's rhetoric, making no public references to the Chinese nuclear program.[33] Yet the possibility of preventive action was not completely abandoned. It seems that McGeorge Bundy was in favor at least until February 1964.[34]

Bundy must have changed his mind somewhere between February and September. On September 15, the top national security leaders – Secretary of State Rusk, Secretary of Defense McNamara, National Security Advisor Bundy, and DCI McCone – met for a lunch to discuss what to do with the Chinese nuclear program. By that time, the Chinese had reached the final stages and the test ground at Lop Nur was being prepared for the first nuclear explosion, whose date the Americans now accurately estimated. Rusk and Bundy agreed that it is better "to have the Chinese test take place" rather than undertake unprovoked unilateral U.S. military action. McNamara concurred but "was non-committal." The position of the usually hawkish McCone is unclear. The four men also decided that close attention should be given to possible pre-emption should the U.S. find themselves in military hostilities with the PRC and that the possibility of joint action with the Soviets, including a cooperative military action, existed and should be discussed with them. Bundy, Rusk, and McNamara then discussed their conclusions with the President, who agreed.[35]

Thus when the Chinese announced the detonation of their first nuclear device on October 16, 1964, President Johnson remained calm, issuing a prepared statement downplaying the test's significance and reassuring allies about the American commitment to their defense.[36] As the Chinese behavior did not become any more threatening, the plans for preventive strikes were slowly wound down. Some residuals survived well into 1965, with respect to possible escalation in Vietnam.[37] The Navy was particularly active in this field, suggesting sinking the Chinese ballistic missile submarine on its maiden voyage and also deploying deterrent patrols of Polaris missile submarines to the Pacific to cover a "new threat target," possibly the counterforce mission.[38] Yet it appears that, by 1966, both countries reached relative satisfaction with the status quo.

Unfolding the complexity

The story of how Kennedy's and Johnson's administrations dealt with the emerging Chinese nuclear arsenal illustrates the complex interplay of various political, military, and normative factors in deterrence. Yet the complexity of deterrence may comprise regularities as well as uniqueness and contingency. To allow for structured comparison across cases and the search for regularities, the following section unfolds this complexity to fit the literature-based theoretical concepts identified in the previous chapter.

The first to be addressed is the deterrer's nuclear arsenal. Even today, with the advantage of hindsight, sources vary in estimating the Chinese nuclear arsenal in the years examined in this case study. It can be said now, without doubt, that the PRC had no functional nuclear device before its first atomic weapon was assembled for the test in Lop Nur in Fall 1964. By that time, the Americans had largely abandoned the idea of a preventive strike targeted at denying China entry into the nuclear club. Yet they also kept their plans for a preventive strike in case the United States should find themselves in a shooting war with Chinese communists. Until Fall 1964, the critical parts of the Chinese nuclear complex were production facilities; most importantly the Baotou Nuclear Fuel Component Plant and the Lanzhou Gaseous Diffusion Plant producing enriched uranium, the plutonium-producing Jiuquan Atomic Energy Complex and the test site in Lop Nur. The size of the arsenal in the months right after the first test was certainly small. Kristensen and Norris estimate that China had some five nuclear weapons in 1965.[39] The availability of delivery vehicles was less than ideal. While Beijing put emphasis on the development of ballistic missiles, those were not available at least until September 1966 and the warhead to fit the missile became first available sometime after a test in October 1966. In the meantime, China had some rudimentary capacity to deliver the nuclear bomb by aircraft, at least after this option was tested in May 1965, yet the design was probably not very successful as the production was canceled.[40] Also the situation of the Chinese air force gave little confidence in using bombers for nuclear weapons delivery as the available bombers had a limited range and were vulnerable to sophisticated U.S. air defense.

The best way to describe the Chinese nuclear arsenal is to use the term "zero to five." This term is well able to capture the uncertainty the challenger must have had about the size and shape of the deterrer's nuclear arsenal. While the Americans did not reportedly think the PRC had any nuclear weapons before 1964, they could not be sure. At the end of the day, the Soviets had promised to deliver a sample atomic weapon to their Chinese counterparts.[41] The deal was secret, but the possibility was hardly left unobserved by the Americans. Before Harriman left for his mission to Moscow, Air Force Chief of Staff General LeMay testified to the Senate Stennis Committee that, at some point, the Soviets might provide China with a nuclear weapon.[42] This possibility must have been somehow considered in Washington to the point that Soviet Foreign Minister Andrei Gromyko felt the need to tell Kennedy on October 10, 1963, that the USSR did not give China "anything."[43] Thus, even before 1964, Americans could not be fully certain that China did not have an atomic bomb, though this possibility was very unlikely. Yet they could be certain that China was somewhere in the zero-to-very-few interval (I use the label of zero to five) and that it had an absolutely rudimentary delivery capability at best.

The Americans may not have been sure about exact state of the Chinese nuclear program but they were certain about the huge disparity between the two. Walt Rostow's comment to Harriman about "U.S. overwhelming nuclear superiority" was not exaggerated.[44] The asymmetry was strong and unquestionable. By 1960, U.S. Air Force Strategic Air Command deployed 1735 B-47 and B-52 jet-powered strategic bombers.[45] By late 1962, the Americans also deployed 284 ICBMs,[46] and from the late 1950s to 1963, U.S. Navy commissioned its first modern SSBNs of the George Washington, Ethan Ellen, and Lafayette classes. By mid-1963, the Navy had 17 ballistic missile submarines with 16 Polaris missiles each.[47] Overall, the number of American nuclear weapons reached 28,133 in 1963, according to Kristensen and Norris, and their destructive power in terms of equivalent megatonnage was only two years behind its historical peak.[48] In the period concerned, the Americans clearly had vast and quite sophisticated nuclear forces, which were able to hit any part of the Chinese territory.

Unless a Waltzian threshold applies (which it is impossible not to pass), China had nothing like a second strike ability. Its nuclear complex was well within the reach of American forces. It could be destroyed even by conventional bombing, though the Chinese air defense would probably cause loses among the attackers. The prospects of survivability from a nuclear strike were even more dismal. Concealment was a viable strategy for survivability improvement, yet the Americans greatly improved their satellite intelligence and had a useful ally in the Chinese nationalists whose intelligence sometimes covered the PRC's nuclear program with impressive accuracy. Furthermore, the available delivery vehicles could not improve China's chance of retaliating even if some of its weapons survived the attack. Ballistic missiles entered service only after the episode, and Chinese ability to deliver the atomic bomb by aircraft was mitigated by the limited range of its bombers and their vulnerability to the sophisticated American air defense.

In general, China's military capabilities could not match America's even in conventional war. Its economy was just overcoming the mess of Mao's Great Leap Forward experiment and was not in a position to support wartime machinery against the world's greatest and most advanced economy. With Soviet help, the Chinese armed forces had undergone substantial modernization from 1954 to 1958 when Soviet organization and tactics were implemented along with Soviet weapons. By the end of 1955, the People's Liberation Army (PLA) had 106 infantry, nine cavalry, 17 artillery, 17 anti-aircraft artillery, and four armored divisions equipped with Soviet weapons. Five thousand Soviet aircraft entered service in its air force (PLAAF) and Soviet ships formed the backbone of the Chinese navy (PLAN). At the same time, Chinese forces reduced their manpower from six to two-and-half million.[49] But this modernization abruptly ended when Mao replaced defense minister Marshal Peng with Lin Biao in 1959.[50] At around the same time, the Soviets terminated their military assistance and left the PRC, at least temporarily, without access to modern military technology. In the following years, China's armed forces grew in numbers to six million, but a substantial part of this number belonged to non-military functions reflecting an increased role of the army in the national economy, while improvements in organization, tactics, and the introduction of state-of-the-art weapons almost stopped.[51] The United States could not compete with the Chinese in terms of numbers. However, it had unequivocal superiority in navy and air force. Against the newborn PLAN stood the numerically and technically preponderant U.S. Navy with more than 20 modern aircraft carriers, a large fleet of surface combatants including some 20 cruisers and more than 200 destroyers, and more than 100 submarines including nuclear-powered ones.[52] A similar situation existed in the air force, though the U.S. preponderance was less in numbers and more in technology. By March 1963, the U.S. Air Force alone had 78 wings organized in 252 squadrons and equipped with 4,504 aircraft, mostly B-47 and B-52 strategic bombers, F-100, F-101, F-102, F-104, F-105, and F-108 interceptor fighters and fighter bombers, KC-97s and KC-105s for strategic air refueling, and various types of transport aircraft.[53] The ground forces were in less favorable conditions vis-à-vis the Chinese. During Eisenhower's years in office, the U.S. Army was reduced to 15 divisions with some 900,000 soldiers, yet it also greatly benefited from increased mobility including aviation, new communication equipment, and new weapons systems.[54] Kennedy and McNamara increased the manpower of U.S. Army to 970,000 soldiers, reactivating two more divisions and allowing existing units to increase their strength.[55] Even with the Marines, reserves, and National Guard units, the numbers were not on the U.S. side, yet its technological preponderance even on the ground was unquestionable. Overall, being greatly inferior in navy, air force, and power projection ability, the PRC was not a military peer to the United States in the global arena. Yet despite unique American capabilities to project power worldwide, the Chinese strength would make it difficult to challenge Beijing in its region.

The limited ability to fight a regional war against the PRC was known to the U.S. military. The Joint Chiefs, whose military expertise was widely respected, had expressed their anxiety that, should the United States strike the PRC's nuclear installations, Chinese communists could in response turn to overt military aggression against South Korea, Taiwan, India, South Vietnam, Cambodia, Laos, Burma, or Thailand. The JCS's estimate was moderately optimistic about U.S. ability to counter any regional military aggression from China, but only if U.S. authorities took timely decisions and were determined to use the force levels required by the situation. The Chiefs clearly expressed that such timely decisions and force levels "would probably include use of tactical nuclear weapons," the need which goes in line with their perception that "the mass use of infantry tactics so widely used by CHICOMS makes their army especially vulnerable to weapons with a high-kill ratio and to chemical, biological, and radiological warfare."[56] As a matter of fact, the Americans had a chance to prove their ability to project power in the region in a few years. While the conflict in Vietnam differs from what would have been required to challenge the PRC's eventual response to strikes on its nuclear installation, some data are revealing. Though the buildup in Vietnam was gradual, while the overt Chinese action discussed by the JCS would have required much faster deployment, the U.S. eventually deployed 545,300 soldiers in the country alone.[57] However, the PRC's force in Vietnam was certainly only a fraction of what it could have deployed and was limited mostly to anti-aircraft artillery divisions, peaking at an impressive 167,000 and certainly causing significant loses among U.S. aircraft.[58] In the rare moments such as NVA's Eastern offensive when United States forces had the opportunity to bring their firepower into action in open battle, the Americans eventually prevailed. But this was against the North Vietnamese, whose resources were limited. It is uncertain that the advantage in firepower would have sufficed to tackle the PLA's superiority in manpower if China brought it to bear. Thus it is probably a safe bet to argue that, in terms of regional conventional preponderance, China was the stronger side, even though its advantage would not last in a longer conflict where the American economy and technology would have a chance to come fully into play.

The U.S.'s strong technological advantage was beyond any doubt. Even Soviet technologies were, on average, considered lagging behind the U.S. ones.[59] Furthermore, the Chinese lost access to state-of-the-art Soviet systems with the Sino-Soviet split in the late 1950s. The PRC managed to produce indigenous clones of Soviet weapons systems that it got before the split to compensate for this loss, but with the rapid pace of improvements in technology during the Cold War, those systems were often quickly becoming obsolete. It must be noted that, when used skillfully, many individual systems such as the J-6 fighter, the Chinese version of MIG-19, proved formidable in Vietnam. Yet despite this performance, most Chinese weapons systems did not match the technological advancement of their U.S. counterparts. The J-6 as the newest PLAAF's fighter was introduced in 1961, but the original MIG-19 came from 1955. It was not until 1963 that China started to produce its first post-Second World War tanks, "Type 59" (Chinese version of the Soviet T-54, which entered service already in the late 1940s), so almost none were in service during the time of this study. Comparable U.S. systems included F-4 Phantoms that entered service in the U.S. Navy in 1960, USAAF's F-105 Thunderchiefs (in service from 1958), and M-60 tanks (in service from 1960). Not only were major Chinese weapons systems of older vintage, at least on average, but China also lacked some powerful systems completely, including intelligence satellites, ballistic missiles, nuclear-powered submarines and aircraft carriers, making the American technological advantage unequivocal.

Despite the valuable advantage technologies like intelligence satellites brought to the United States, the information available to Washington was always somewhat limited. This was understandable in the first years of the case study. The U-2 flights were difficult due to the remote location of China's installations and CORONA satellites had a limited resolution. Yet, by 1963, at the latest, CORONA's resolution had improved, new GAMBIT satellites had been launched, and U-2 flights were arranged from Indian airfields and from Taiwan.[60] However, even with improvements in available coverage, some estimates remained flawed. The United States rightly identified critical installations including Lanzhou, Lop Nur, and others, yet up to the moment when samples from the first Chinese nuclear explosion were collected, the U.S. intelligence supposed that the PRC was pursuing a plutonium-based, not a uranium-based, track to the bomb.[61] This omission is interesting. Washington received the necessary information to reassess PRC's uranium enrichment program from Taiwanese intelligence, but the information was probably disregarded.[62] It appears that the Chinese nationalists had an espionage network of substantial value on the mainland that was even able to provide useful information on the PRC's nuclear issues that Beijing kept in complete secrecy, but the Americans did not fully trust Chiang's government.[63]

The unfolding perception of China by American decision makers is at least as important as the unfolding military factors above. Clearly, the conflict with China was perceived through the lens of the Cold War. Yet Kennedy's and Johnson's administrations were very sensitive to the Sino-Soviet split and able to modify their perceptions of the Cold War accordingly. While some authors argue that Vietnam, Laos, Berlin, and Cuba were more central to Kennedy than China's nuclear program, Kennedy himself vocally expressed the fact that nuclear China was high on his list of threats to American interests.[64] Though it is unquestionable that the Soviet Union was perceived as the main adversary, it must also be noted that the Soviets appeared to Kennedy as less dangerous and, importantly, as more predictable. Kennedy was even willing to accept some Soviet cheating under the cover of the test ban treaty if it meant stopping the PRC's nuclear ambitions.[65] Thus Kennedy's perception was not disassociated from the Cold War logic. Rather, he believed that if the PRC got the bomb, the whole South-East Asia would fall into communist hands.[66] The fact that President Kennedy discussed the matter with Chiang during his visit to Washington also suggests that the PRC's nuclear program was not of marginal importance to him. According to Xia, JFK felt that the Chinese nuclear program was probably the most serious problem facing the world.[67]

This perception of threat was closely linked to the perception of Mao's strong resolve. The Chinese communists had proven to be able to accept considerable hardships during the war against Chiang's nationalist forces and the Japanese occupation army. The PRC had also established its reputation through its resolute intervention into the Korean War. Beijing intervened despite nuclear threats from United States, and while Beijing may not have considered the nuclear threats credible, it was surely aware of its grave situation after the bloody civil war.[68] Furthermore, in Korea, Beijing had once again shown remarkable tolerance of its own causalities, losing 200,000 soldiers in only the first six weeks of the war and yet remaining in combat.[69] This track record of Chinese resolve could not be overlooked and Mao publicly voiced his disrespect of American threats. He made numerous references to being ready to fight under nuclear conditions. In November 1957, during a meeting of communist parties in Moscow, he even urged others that they should not be frightened by the prospect of a nuclear war started by imperialists as it would be costly but would destroy the capitalist system.[70] While Mao's rhetoric probably bears a degree of bluff, his country was certainly not viewed as an irresolute one that would surrender to American intervention.

A correct estimate of the alleged Chinese threat was difficult for the Americans. Despite fierce Cold War rhetoric, for the better part of the conflict the Soviet Union and the United States enjoyed a rich set of formal and informal diplomatic ties that enabled communication between them. On the contrary, trade and normal diplomatic relations did not exist between Beijing and Washington.[71] China was not even admitted to the UN, and, until 1966, U.S. diplomats were not allowed to make informal social contacts with Chinese Communists.[72] Diplomatic channels were mostly limited to ambassadorial talks in Warsaw, where 16 meetings were held during Kennedy's term between PRC Ambassador Wang Bingnam and U.S. Ambassadors Jacob Beam (till November 1961) and John Cabot. However, for most of the time, nothing was really discussed as both sides kept presenting their rigid positions.[73] In fact, the institutionalization of mutual relations was remarkably limited and the differences in the two actors' perceptions remarkably strong. Even these days, both sides read each other's military signals with a considerable degree of misunderstanding, and this was even more valid in the 1950s and 1960s.[74]

This lack of institutionalization goes hand-in-hand with a history of hostility between the United States and the Chinese communists. U.S. support for Chiang Kai-shek's nationalists during the Chinese civil war could not keep him in power over the mainland, but well sufficed to alienate future relations with Mao, while the shadows of Truman's "loss of China" made any future democratic American president careful not to appear soft on the Chinese communists.[75] Further, the strong anti-communist mood that prevailed in the United States in the early 1950s prevented the establishment of relations between Washington and the new government in Beijing, particularly when China entered into the alliance with Stalin. The ideological differences appeared impenetrable and the hostilities escalated to a new level with the outbreak of the Korean War, only to advance even more with the Chinese intervention.[76] With the Korean War, the United States turned their largely vocal support of the GRC into a real military involvement. Washington deployed its Seventh Fleet in the region and committed itself to the defense of Taiwan in the Formosa resolution.[77]

The recent history the Americans had with the Chinese communists made leaders in Washington question Beijing's rationality. The Americans believed that China would be ready to sacrifice hundreds of millions.[78] The willingness to making such a sacrifice could hardly appear rational to U.S. decision makers. It is, then, little surprising that a National Intelligence Estimate from December 6, 1960, denounced PRC's "arrogant self-confidence, revolutionary fervor, and distorted view of the world" that "may lead to miscalculate risks" and further that "this danger would be heightened if Communist China achieved a nuclear weapons capability."[79] This view goes in line with Kennedy's later instruction to Harriman for his mission to negotiate the test ban in Moscow that "small forces in hands of people like CHICOMS could be very dangerous."[80] Certainly not all Americans shared this view of China. Herbert Johnson of the State Department led the reassessment of China's nuclear threat, which was established on a different perception of Beijing's taste for aggression and rationality. But while Herbert Johnson's analyses may have directly or indirectly influenced President Johnson's decision not to strike China before it tested its first atomic bomb, he certainly did not influence his predecessor, who appeared to see a strike against China's nuclear facilities almost as the last resort, arguing that "we can't afford to let them do this."[81]

Turning to international factors, it does not appear that these played a major role, apart from the American attempts to convince the Soviets to join the anti-Chinese action. The JCS recommended getting the USSR on board for a joint action, considering it as a possible tipping point between Chinese resistance and submission, but did not expect the USSR to accept the American attempt.[82] Washington certainly preferred to have the Soviets join, or at least approve American steps, but the Soviets refused to discuss the issue, giving mostly equivocal responses to U.S. probes. Yet it also does not seem that Moscow actively played the role of a third-party deterrer. The closest the Soviets probably got to directly dissuading possible U.S. intervention was when Ambassador Dobrynin reminded Secretary of State Rusk of the existence of a mutual defense treaty from February 1950.[83] However, by that time, the treaty was largely a piece of paper and it is uncertain whether the Americans considered the threat of Soviet involvement credible. Washington was sufficiently informed about the Sino-Soviet split. CIA reported as of early 1961 that the rift between Moscow and Beijing was "genuine, serious and bitter."[84] Thus, on September 18, 1964, less than a month before China's first nuclear test, NSC staffer Robert Kromer wrote to National Security Advisor Bundy that should the United States destroy the Chinese nuclear facilities "the Soviets would approve privately, but might have to raise a to-do publicly."[85]

Similarly to third-party deterrence, alliance politics could probably only marginally influence the American decision not to strike. The possible impact of the Chinese nuclear explosion on U.S. allies was considered in Washington, yet if anything, such consideration only tempted the Americans to strike. As Lawrence Freedman observed, "Kennedy's instinct ... was to work with allies he found congenial."[86] From the long list of American allies, he certainly discussed the PRC's nuclear program with the British and the Taiwanese. British Prime Minister Harold Macmillan shared Kennedy's concerns about the Chinese nuclear program and both leaders discussed methods for preventing proliferation with respect to China.[87] Chiang's eagerness to have his archenemy's nuclear program destroyed was only too obvious. And, similarly, the Japanese as the key allies in Asia were deeply anxious about a Chinese nuclear test and asked the United States to be ready to use nuclear weapons if a war occurred between China and Japan.[88]

While the influential allies were generally supportive, international legitimacy of action was somewhat more dubious. The problem was not thoroughly discussed in Washington, indicating a rather limited importance, but the Americans were certainly aware of the nonintervention norm. Robert Kromer touched upon this issue in his lengthy memorandum to Bundy. Kromer argued that "in the rest of the world there would be considerable fear – also some feeling that the U.S. was punishing a smaller power for getting into the nuclear business." But continuing, "was this necessarily bad, however?" he questioned the negative consequences of such feelings in the world, and downplayed the negative consequences further by arguing that "initial fears might quickly turn to relief once the crisis seemed to pass."[89] The Joint Chiefs were rather more skeptical, arguing that a small-scale conventional attack would be "difficult to justify to world opinion" and that the eventual use of tactical nuclear weapons "would expose United States to strong criticism from world opinion." Nonetheless, at the end of the day, the JCS recommended using tactical nuclear weapons for the mission despite the expected negative consequences on world opinion, suggesting the action's legitimacy.[90]

The JCS recommendation to use tactical nuclear weapons was probably met with an unenthusiastic reaction from civilian leaders. According to Tannenwald, the Kennedy administration was slowly internalizing the normative prohibition of the first use of nuclear weapons, even though the norm was still far from strong.[91] Certainly political leadership did not consider using tactical nuclear weapons against the Chinese arsenal as a preferable option and, despite reported military advantages, ordered the JCS to plan for a conventional strike.[92] However, it seems that this decision was rather driven by the logic of consequences than by the logic of appropriateness. The administration expected a more moderate response from China and world opinion should the conventional weapons handle the mission, rather than refused the nuclear option on strong normative grounds.

Though with only a limited impact on Kennedy's and Johnson's decisions, the international legitimacy of the action was probably more constraining than the domestic one. Acting against the Chinese communists largely appeared legitimate to the American elite, and anti-PRC views were remarkably strong among American people, imposing little constraint on possible action. In a rare interference into his successor's business, President Eisenhower warned that he "would feel it necessary to return to political life if the Chinese communists were admitted to the United Nations."[93] His position was not an unusual one. In a 1963 poll, 47 percent of Americans considered the PRC to be a bigger threat than the Soviet Union, while only 34 percent had the opposite view.[94] An influential pro-nationalist "China Lobby" in Washington at least partially managed to shape the U.S. position on PRC.[95] Thus, surgical bombing of China's nuclear installation would have probably been met with little opposition. The republicans at the time were generally even more hostile to communist China, nominating hawkish Barry Goldwater as their presidential candidate. The democrats had to live with the history of "losing China" and carefully watched not to appear weak on Beijing. As Burr and Richelson suggest, President Johnson may have been limited in his ability to take military action while he was running for re-election against Goldwater on a "peace platform."[96] Yet the evidence is inconclusive and it is unlikely that dovish voters would have switched to Goldwater had Johnson opted for the strike.

Fiercer opposition to the strike was located in the civil service. There, starting sometime in 1963, Robert Johnson from the State Department's Policy Planning Staff led the reassessment of the impact of China's nuclearization on the United States. Johnson's arguments managed to convince his director Walt Rostow to pass his study up the chain of command. Secretary of State Dean Rusk "was favorably impressed" by Johnson's study and it appears that from October 1963 the State Department's position largely reflected his conclusion.[97] Nevertheless it must be also noted that the State Department's position may not have been decisive. Other influential players did not necessarily share its feeling that nuclear weapons at the hands of the PRC were acceptable, and the CIA and Pentagon kept planning possible ways to destroy the Chinese nuclear program for some time to come.

Considering the factor of regime type, there is little doubt that, despite some imperfections, the United States in the 1960s can be analytically treated as a full democracy.[98] The role of the military in this regime is only marginally more difficult to access. As an experienced top-ranking soldier, President Eisenhower knew all the tips and tricks used by U.S. military to influence civilian leaders. Perhaps counterintuitively to some observers, the former general did not promote the military's interest, but rather pursued a policy of strict control over the military, whose politics he understood only too well. This policy was codified in the Department of Defense Reorganization Act of 1958 that enhanced the independence of the JCS and reduced the role of services. The Kennedy administration, namely Secretary McNamara, took full advantage of the new legislation and subjected the military to an elaborate control of civilians, now interfering even into procurements and other affairs services had deemed as their exclusive prerogative before.[99] Thus, while relations between military and civilian leadership were occasionally tense, the civilians were decisively in control.

Perhaps most interesting, particularly with respect to the analytical primacy of nuclear deterrence, is to unfold the American reflection of China's military deterrence vis-à-vis an American preventive strike. With respect to the present U.S.–China case, such analytical primacy looks completely unreasonable. Washington was gravely concerned about the possible political and psychological impact of nuclear-armed communist China, but the military factor was less disturbing. Contrary to the views of minimalists, existentialists, and the like, it was largely believed than nuclear superiority does matter. According to Burr and Richelson, a few weeks before Harriman left for Moscow, Walt Rostow told him that the "minimal nuclear capability that Beijing could develop was unlikely to convince … anyone" and that "U.S. overwhelming nuclear superiority" would deter China.[100] What is even more telling about the American perception of Chinese nuclear deterrence is the apparent lack of references to the threat of China's nuclear retaliation and the lack of planning to handle such event in the top-secret military materials like the JCS response to Nitze's request for a military options study. This was in visible contrast with the Joint Chiefs' explicit concerns about China's conventional actions that can be taken in response to U.S. attack against its nuclear installations. It also does not appear that such a perception was limited only to the embryotic stage of China's nuclear weapons development. Rather it seems that after China's nuclear test demonstrated its nuclear capability the United States kept planning for a preventive strike should the need arise and took steps to facilitate options, such as the deployment of counterforce-capable Polaris submarines into the Pacific. Evidence from between the time of China's nuclear test in 1964 and 1966, when the situation stabilized, does not suggest enhanced fear of China's possible nuclear retaliation in Washington.

Contrary to the threat of nuclear retaliation, the threat of conventional retaliation was clearly in the minds of American decision makers. American allies in Asia, particularly South Korea, South Vietnam, Taiwan, and also Laos, Burma, Cambodia, Thailand, and India appeared vulnerable to China's covert or overt military action. The credibility of China's conventional threat was unequivocal. The wars China had waged in Korea and on the Sino-Indian borders and the repeated clashes in the Taiwan Strait were a living memory. Secretary of State Dean Rusk was afraid that China would intervene with enormous manpower,[101] and he was hardly the only one with such an anxiety. As the hostilities in Vietnam grew, the Johnson administration had a major fear that China's intervention would turn Vietnam from a regional war into something much more worrying.[102] While the JCS was optimistic about U.S. ability to tackle China's conventional aggression by proper reaction and timely deployment, there was a major "but." The Chiefs clearly considered a tactical nuclear response against China's conventional aggression necessary, yet such a step appeared incompatible with the emerging and growing taboo of the first use. Yet the Korean experience demonstrated only too well that China's manpower was difficult to defeat with conventional weapons only.

China's nuclear denial can be safely handled by bluntly saying that it simple did not play any role, but conventional denial deserves more elaboration. The PRC was well aware that Washington would be unenthusiastic about its nuclear program and might contemplate a preventive attack. Thus, from the beginning, Beijing took careful precautions. The program was run in secrecy and critical parts were located in remote regions of China. Furthermore, all installations had significant anti-aircraft protection, making it more difficult to use manned bombers, which were the most effective counterforce weapon of the time and the only one which could have handled the mission with conventional munitions.[103] While conventional bombing was the preferable option for U.S. civilian leaders, it would have required many sorties.[104] China's anti-aircraft defense would have likely claimed its costs on the attacking aircraft. Yet such costs did not appear prohibitive, and the conventional mission was deemed feasible.[105]

Table 2.1 summarizes the empirical results of this case. Most notably, it shows that the United States was little concerned by the possibility of China's eventual nuclear retaliation even though Washington was less then 100 percent certain that China had not obtained a sample bomb from the Soviet Union. On the contrary, leaders in Washington seriously debated military action against China's nuclear program. Many of them felt compelled to respond before the perceived Chinese nuclear threat fully materialized, because they saw the regime in Beijing as being hostile and irrational. Yet several factors raised important doubts in the heads of those who considered striking. Most important of them was a fear of China's conventional retaliation in Asia. There, in an eventual war, the United States ground forces would have been the conventionally weaker party, which could only defeat PLA's huge numbers by battlefield use of tactical nuclear weapons, an eventuality that appeared unpleasant under the logic of consequences.

Table 2.1 U.S.–China, 1959–1969

Concept	Description	Values
The deterrer's nuclear arsenal	Key physical attributes of respective nuclear complexes; weapon types; numbers; delivery vehicles; command, control, and communication systems	0–5 (possibly some crude nuclear weapons; production facilities; no early warning system; primitive command and control)
Nuclear asymmetry	Situation when a substantially smaller (quantitatively) and less sophisticated nuclear force (qualitatively) faces a qualitatively and quantitatively larger nuclear force	Strong asymmetry (quantitative superiority by a factor of approx. 1,000 and a comparable qualitative edge)
Second-strike criterion	Nuclear posture that has the ability to survive enemy attack, make and communicate decision to retaliate, overcome enemy's active defense, and destroy a valuable target despite its passive defense	None (loopholes in survivability; imperfect C3 systems; inadequate delivery vehicles available)
General conventional preponderance	Situation when one side's armed forces are in general substantially stronger in terms of numbers, technology, training, and employment strategies	Challenger (stronger navy and air force; numerically weaker land forces but a substantial edge in technology and firepower per unit)
Theater conventional preponderance	Situation when one side's armed forces are substantially stronger in terms of numbers, technology, training, and employment strategies on a theater where targets valuable to the challenger can be found	Deterrer (immediate superiority; likely the challenger's superiority after full mobilization in the long term)
The challenger's technological advantage	Significant advantage of the challenger's major weapons systems that would be employed in case of conflict in terms of state-of-the-art sophistication over the deterrer's weapons systems	Strong (significant edge in the sophistication of most weapon platforms; some technologies completely unavailable to the deterrer)

Availability of information	The challenger's knowledge about targets' location and defensive systems, which is established from sources that the challenger deems credible	Limited (inadequacy in imagery; ineffective utilization of GRC information; but good knowledge of production complexes location)
Centrality of conflict	Absolute importance of the conflict dyad between the challenger and the deterrer from the challenger's subjective perspective and its relative importance to other existing conflict dyads where the challenger is a party	Yes (attention to the conflict paid by the President himself; conflict seen as part of a central cold war confrontation with communism)
Perceived resolve of the deterrer	The challenger's perception of the deterrer's commitment to fight should the deterrence fail	Strong (demonstrated during wars with Japan, nationalists, and in Korea; vocal disrespect of U.S. nuclear threats)
Institutionalization of mutual relations	Degree of shared expectations about the requirements of stable deterrence and the existence of proven formal or informal communication channels	Limited (contacts limited to occasional ambassadorial talks, otherwise virtually no diplomatic relations or trade; PRC not even in the UN)
History of hostility	Track record of conflict in the dyad that shapes the understanding of other side's intentions	Strong (U.S. support to Chinese nationalists; war in Korea; perception of conflict through the anti-communist lens)
Last resort	Situation when the challenger sees only the options to strike, or to live up to the development he tries to prevent	Strong/limited (strong with JFK; declining after his death with a reassessment authored by the State Department)
Perception of the deterrer's rationality	The challenger's perception of the deterrer's rationality, particularly whether the challenger believes it possible to live with the nuclear-armed deterrer in the long term	No (PRC viewed as largely irrational by JFK; perception changing after his death with a reassessment authored by the State Department)

continued

Table 2.1 Continued

Concept	Description	Values
Third-party deterrence	Threat of military involvement into the original conflict by a third party, most likely the allies of the original deterrer, that decisively influenced the challenger's decision	Limited (USSR commitment to come to help PRC no longer considered credible given the Sino-Soviet split)
Alliance politics	Sensitivity of the challenger to possible impact of his action on the relations with his allies	None (no signs of allied opposition to action; Taiwan even suggested doing the job for the U.S.)
International action legitimacy	The challenger's sensitivity to the international normative expectation of non-intervention	Limited (concerns about action legitimacy raised but not considered prohibitive)
Nuclear taboo	Normative prohibition on the use of nuclear weapons	Emerging (nuclear use suggested by the JCS overruled by civilians, but largely driven by the logic of consequences)
Domestic action legitimacy	Level of support for military solution among the challenger's population	Strong (the PRC viewed as a bigger threat than the USSR; strong hawks and anti-China lobby)
Opposition among influential decision makers	Negative view on military solution by influential part of the challenger's government	Strong (led by the State Department)

Regime type	The challenger's position on the democracy-non-democracy axis and on the militarized-non-militarized axis	Democracy, strong civilian control of armed forces
Nuclear retaliation	Threat that the deterrer will use its nuclear weapons against targets valuable to the challenger, except of targets that are directly related to pursuit of the challenger's objectives	Limited (no concerns about China's nuclear retaliation raised; no preparatory measures taken to mitigate such threat)
Nuclear denial	Threat that the deterrer will use its nuclear weapons against targets that are directly related to pursuit of the challenger's objectives in order to prevent him from attaining the objectives, or in order to make it unacceptably costly	Limited (no concerns about China's nuclear denial raised; no preparatory measures taken to mitigate such threat)
Conventional denial	Threat that the deterrer will use its conventional weapons against targets that are directly related to pursuit of the challenger's objectives in order to prevent him from attaining the objectives, or in order to make it unacceptably costly	Limited (some concerns about China's anti-aircraft protection of targets raised, but not considered prohibitive; tactical nuclear weapons suggested as useful in minimizing the number of sorties needed)
Conventional retaliation	Threat that the deterrer will use its conventional weapons against targets valuable to the challenger, except of targets that are directly related to pursuit of the challenger's objectives	Strong (concerns about China's covert or overt action against U.S. allies in Asia; conventional buildup and tactical nuclear response suggested by JCS)

Notes

1 John W. Lewis and Xue Litai, *China Builds the Bomb* (Stanford: Stanford University Press, 1991) 35–39; Jeffrey Lewis, *The Minimum Means of Reprisal: China's Search for Security in Nuclear Age* (Cambridge: MIT Press, 2007) 55–56.

2 Xiaobing Li, *History of the Modern Chinese Army* (Lexington: University Press of Kentucky, 2007) 149.

3 Therese Delpech, *Nuclear Deterrence in the 21st Century: Lessons from the Cold War for the New Era of Strategic Piracy* (Santa Monica: RAND, 2012) 70–73.

4 William Burr and Jeffrey T. Richelson, "Whether to 'Strangle the Baby in the Cradle': The United States and the Chinese Nuclear Program, 1960–1964," *International Security*, 25/3 (Winter 2000/2001) 58.

5 Lewis and Xua, *China Builds the Bomb*, 115

6 Li, *History of the Modern Chinese Army*, 149.

7 Zhihua Shen and Yafeng Xia, "Between Aid and Restriction: The Soviet Union's Changing Policies on China's Nuclear Weapons Program 1954–1960," *Asian Perspective* 36/1 (April 2009) 66–112.

8 Jeffrey Lewis, *Minimum Means of Reprisal*, 56.

9 Lyle J. Goldstein, *Preventive Attack and Weapons of Mass Destruction: A Comparative Historical Analysis* (Stanford: Stanford University Press, 2006) 57.

10 "Memorandum of Discussion at the 448th Meeting of the National Security Council," June 22, 1960, *FRUS, 1958–1960*, volume XIX, China, document 338.

11 "NIE 13-2-60: The Communist Chinese Atomic Energy Program," December 13, 1960, *FRUS, 1958–1960*, volume XIX, China, document 364.

12 Gordon Chang, "JFK, China, and the Bomb," *The Journal of American History*, 74/4 (March, 1988) 1288.

13 Chang, "JFK, China, and the Bomb," 1289–1290.

14 Goldstein, *Preventive Attack and Weapons of Mass Destruction*, 57.

15 Chang, "JFK, China and the Bomb," 1291.

16 See Lawrence Freedman, *Kennedy's Wars: Berlin, Cuba, Laos, and Vietnam* (Cary: Oxford University Press, 2000) Chapters 28, 29.

17 Chang, "JFK, China and the Bomb," 1300.

18 Freedman, *Kennedy's Wars*, 272–274.

19 Chang, "JFK, China and the Bomb," 1300–1304; Burr and Richelson, "Whether to 'Strangle Baby in the Cradle'," 71.

20 Burr and Richelson, "Whether to 'Strangle Baby in the Cradle'," 64.

21 Burr and Richelson, "Whether to 'Strangle Baby in the Cradle'," 68.

22 RG 59, Records of Bureau of Far Eastern Affairs, Office of the Country Director for the Republic of China, Top Secret Files Relating to the Republic of China, 1954–65, box 4, 1963 Top Secret Nuclear Capability Briefing Book on U.S.–Soviet Non-Diffusion Agreement. For discussion at Moscow meeting, General Curtis E. LeMay, Acting Chairman, Joint Chiefs of Staff, to Secretary of Defense, "Study of Chinese Communist Vulnerability," April 29, 1963.

23 "Telegram from the Embassy in the Soviet Union to the Department of State," July 27, 1963, *FRUS, 1961–1963*, volume VII, Arms Control and Disarmament, document 354.

24 "Memorandum from Robert W. Komer of the National Security Council Staff to the President's Special Assistant for National Security Affairs," February 26, 1964, *FRUS, 1964–1968*, volume XXX, China, document 14, fn. 7.

25 "Memorandum from Robert W. Komer of the National Security Council Staff to the President's Special Assistant for National Security Affairs," February 26, 1964, *FRUS, 1964–1968*, volume XXX, China, document 14, fn. 7.

26 "Memorandum from the Joint Chiefs of Staff to the Secretary of State," March 2, 1964, FRUS, 1964–1968, volume I, Vietnam, 1964, document 66.

27 "Meeting Between Mr. McGeorge Bundy and General Chiang Ching-kuo," September 10, 1963, *FRUS, 1961–1963*, volume XXII, Northeast Asia, document 185.

28 "Memorandum of Conversation," September 11, 1963, *FRUS, 1961–1963*, volume XXII, Northeast Asia, document 186.

29 Parts of Chiang's response to the President in "Memorandum of Conversation" remain classified, which is one of the only two parts of this lengthy document not yet declassified. The other one is a part of Chiang's response to Kennedy's question about the number of GRC's intelligence agents on the mainland.

30 Burr and Richelson, "Whether to 'Strangle Baby in the Cradle'," 73.

31 Robert H. Johnson, State Department Policy Planning Council, "A Chinese Communist Nuclear Detonation and Nuclear Capability: Major Conclusions and Key Issues," October 15, 1963, Policy Planning Council Records, 1963–64, box 275, S/P Papers Chicom Nuclear Detonation and Nuclear Capability, Policy Planning Statement, 10/15/63.

32 Burr and Richelson, "Whether to 'Strangle Baby in the Cradle'," 76.

33 Burr and Richelson, "Whether to 'Strangle Baby in the Cradle'," 79.

34 Goldstein, *Preventive Attack and Weapons of Mass Destruction*, 61.

35 This is based on McCone's and Bundy's recalls of the meeting from the "Memorandum for the Record," September 15, 1964, FRUS, 1964–1968, volume XXX, China, document 49; and "Memorandum for the Record," September 15, 1964, FRUS, 1964–1968, volume XXX, China, document 50.

36 Burr and Richelson, "Whether to 'Strangle Baby in the Cradle'," 92.

37 "Information Memorandum from the Acting Deputy Under Secretary of State for Political Affairs to Secretary of State Rusk," July 15,1965, *FRUS, 1964–1968*, volume XXX, China, document 94.

38 Goldstein, *Preventive Attack and Weapons of Mass Destruction*, 63, 65–66.

39 Hans M. Kristensen and Robert S. Norris, "Global Nuclear Weapons Inventories, 1945–2013," *Bulletin of the Atomic Scientists*, 69/5 (2013) 78.

40 Lewis, *Minimum Means of Reprisal*, 62–63.

41 Burr and Richelson, "Whether to 'Strangle Baby in the Cradle'," 58.

42 Chang, "JFK, China, and the Bomb," 1306.

43 Burr and Richelson, "Whether to 'Strangle Baby in the Cradle'," 76.

44 Burr and Richelson, "Whether to 'Strangle Baby in the Cradle'," 76.

45 Gareth Porter, *Perils of Dominance: Imbalances of Power and the Road to Vietnam* (Berkeley: University of California Press, 2005) 5.

46 Porter, *Perils of Dominance*, 7.

47 James L. Mooney *et al.*, *Dictionary of American Naval Fighting Ships* (available online at www.history.navy.mil/danfs/index.html).

48 Kristensen and Norris, "Global Nuclear Weapons Inventories, 1945–2013," 78; Albert Wohlstetter, "Racing Forward? Or Ambling Back?," in Robert Zarate and Henri Sokolski (eds), *Nuclear Heuristics: Selected Writings of Albert and Roberta Wohlstetter* (Carlisle: Strategic Studies Institute, U.S. Army War College, 2009) 442; Equivalent megatonnage is a standardized measure to compare the destructive power of nuclear arsenals. This destructive power is measured as the area to be destroyed by the arsenal as a multiple of the area that

would be destroyed by a one-megaton weapon. After the 1960s, improvement in weapon accuracy allowed the yield of American nuclear weapons to decline.

49 Li, *History of the Modern Chinese Army*, 119–125.
50 Li, *History of the Modern Chinese Army*, 178.
51 Li, *History of the Modern Chinese Army*, 198.
52 James L. Mooney *et al.*, *Dictionary of American Naval Fighting Ships* (available online at www.history.navy.mil/danfs/index.html).
53 USAF, *United States Air Force Statistical Digest: Fiscal Year 1963* (Washington D.C.: Directorate of Data Automation of the Air Force Headquarters, 1963) 15. Additional aircraft of various types, including fighters and ground attack tactical bombers, were operated by U.S. Navy and several thousand additional fixed-wings aircraft and helicopters belonged to the U.S. Army.
54 Richard W. Steward *et al.*, *American Military History,* volume II: *The United States Army in a Global Era, 1917–2008* (Washington D.C.: Center of Military History U.S. Army, 2010) 262.
55 Steward *et al.*, *American Military History,* volume II, 277–288.
56 RG 59, Records of Bureau of Far Eastern Affairs, Office of the Country Director for the Republic of China, Top Secret Files Relating to the Republic of China, 1954–65, box 4, 1963 Top Secret Nuclear Capability Briefing Book on U.S.–Soviet Non-Diffusion Agreement for Discussion at Moscow Meeting, General Curtis E. LeMay, Acting Chairman, Joint Chiefs of Staff, to Secretary of Defense, "Study of Chinese Communist Vulnerability," April 29, 1963.
57 Steward *et al.*, *American Military History,* volume II, 346.
58 Li, *History of Modern Chinese Army*, 217–219.
59 Porter, *Perils of Dominance*, 3.
60 Burr and Richelson, "Whether to 'Strangle Baby in the Cradle'," 63, 84.
61 Burr and Richelson, "Whether to 'Strangle Baby in the Cradle'," 56.
62 Burr and Richelson, "Whether to 'Strangle Baby in the Cradle'," 65.
63 See fn. 27.
64 Chang, "JFK, China, and the Bomb," 1288.
65 Goldstein, *Preventive Attack and Weapons of Mass Destruction*, 58.
66 Goldstein, *Preventive Attack and Weapons of Mass Destruction*, 57.
67 Yafeng Xia, *Negotiating with the Enemy: U.S.–China Talks during the Cold War, 1949–1972* (Bloomington: Indiana University Press, 2006) 108.
68 Delpech, *Nuclear Deterrence in the 21st Century*, 70–71.
69 Christopher P. Twomey, *Military Lens: Doctrinal Differences and Deterrence Failure in Sino-American Relations* (Ithaca: Cornell University Press, 2010) 157.
70 Li, *History of the Modern Chinese Army*, 155.
71 Chang, "JFK, China, and the Bomb," 1300.
72 Xia, *Negotiating with the Enemy*, 122.
73 Xia, *Negotiating with the Enemy*, 106–134.
74 See Twomley, *Military Lens*.
75 Roger Buckley, *United States in the Asia-Pacific since 1945* (West Nyack: Cambridge University Press, 2002) 58–63.
76 Robert Jervis, "The Impact of the Korean War on the Cold War," *The Journal of Conflict Resolution*, 24/4 (December, 1980) 563–592.
77 Edward J. Marolda, *Ready Seapower: A History of U.S. Seventh Fleet* (Washington D.C.: Naval History & Heritage Command, 2012) 40–44.
78 Goldstein, *Preventive Attack and Weapons of Mass Destruction*, 58.
79 "National Intelligence Estimate," December 6, 1960, *FRUS, 1958–1960*, volume XIX, China, document 362.

80 Fn. 15.

81 Goldstein, *Preventive Attack and Weapons of Mass Destruction*, 58.

82 RG 59, Records of Bureau of Far Eastern Affairs, Office of the Country Director for the Republic of China, Top Secret Files Relating to the Republic of China, 1954–65, box 4, 1963 Top Secret Nuclear Capability Briefing Book on U.S.–Soviet Non-Diffusion Agreement for Discussion at Moscow Meeting, General Curtis E. LeMay, Acting Chairman, Joint Chiefs of Staff, to Secretary of Defense, "Study of Chinese Communist Vulnerability," April 29, 1963.

83 Goldstein, *Preventive Attack and Weapons of Mass Destruction*, 70.

84 Freedman, *Kennedy's Wars*, 257.

85 "Memorandum From Robert W. Komer of the National Security Council Staff to the President's Special Assistant for National Security Affairs," *FRUS, 1964–1968*, volume XXX, China, document 51.

86 Freedman, *Kennedy's Wars*, 277.

87 Freedman, *Kennedy's Wars*, 265; Burr and Richelson, "Whether to 'Strangle Baby in the Cradle'," 70.

88 Delpech, *Nuclear Deterrence in the 21st Century*, 77.

89 Fn. 81.

90 RG 59, Records of Bureau of Far Eastern Affairs, Office of the Country Director for the Republic of China, Top Secret Files Relating to the Republic of China, 1954–65, box 4, 1963 Top Secret Nuclear Capability Briefing Book on U.S.–Soviet Non-Diffusion Agreement for Discussion at Moscow Meeting, General Curtis E. LeMay, Acting Chairman, Joint Chiefs of Staff, to Secretary of Defense, "Study of Chinese Communist Vulnerability," April 29, 1963; and "Memorandum from Robert W. Komer of the National Security Council Staff to the President's Special Assistant for National Security Affairs," February 26, 1964, *FRUS, 1964–1968*, volume XXX, China, document 14, fn. 7.

91 Nina Tannenwald, *The Nuclear Taboo: The United States and the Non-Use of Nuclear Weapons Since 1945* (Cambridge: Cambridge University Press, 2007) 251–261.

92 Goldstein, *Preventive Attack and Weapons of Mass Destruction*, 68.

93 Freedman, *Kennedy's Wars*, 250.

94 Freedman, *Kennedy's Wars*, 249.

95 Stanley D. Bachrack, *The Committee of One Million: "China Lobby" Politics, 1953–1971* (New York: Columbia University Press, 1976).

96 Burr and Richelson, "Whether to 'Strangle Baby in the Cradle'," 88.

97 Burr and Richelson, "Whether to 'Strangle Baby in the Cradle'," 77–79.

98 This corresponds with the ranking the U.S. received in the widely respected "Polity Database IV."

99 Eric R. Mahan, "Civil–Military Relations" in Marc J. Selverstone (ed.) *A Companion to John F. Kennedy* (Chicester: Whiley-Blackwell, 2014) 176–180.

100 Burr and Richelson, "Whether to 'Strangle Baby in the Cradle'," 76.

101 Goldstein, *Preventive Attack and Weapons of Mass Destruction*, 71.

102 Li, *History of Modern Chinese Army*, 216.

103 Both ICBMs and SLBMs were largely inaccurate at the time and required nuclear warheads of substantial yield based on the degree of hardening of the target to compensate for this inaccuracy.

104 Burr and Richelson, "Whether to 'Strangle Baby in the Cradle'," 80.

105 Goldstein, *Preventive Attack and Weapons of Mass Destruction*, 67.

3   The Soviet Union and China, 1969

The small border war between the Soviet Union and China that took place in 1969 reportedly brought the Cold War to an almost forgotten equivalent of the Cuban Missile Crisis. Yet, happening far from civilization, at an unpopulated border between two secretive regimes that have so far kept much of their valuable historical data locked in the archives, the events mostly avoided larger scholarly attention; a striking contrast to the 1962 crisis at the Caribbean that is now one of the most carefully examined single events in nuclear history. The few researchers who studied the crisis, which allegedly made leaders in Moscow consider striking at the young Chinese nuclear arsenal, struggle with a lack of available archival sources. To make matters even more problematic, more is known about Beijing's fears of the Soviet strike than about Moscow's actual planning. The lack of evidence makes it even possible that the Soviet nuclear threats were only an elaborate bluff. Yet the importance of this case for the theory more than justifies studying it, even under imperfect conditions. In 1969, the Chinese nuclear arsenal was small, unsophisticated yet surely existing. That makes the crisis one of the few that can reveal observable mechanisms of nuclear deterrence between the small and the big. With such a small number of empirically available cases, students of nuclear deterrence cannot be too picky in their case selection.

The plot

The 1969 events were the most dangerous byproduct of the Sino-Soviet split. They grew from the complicated and complex relations between the two communist parties which dated back to 1920s. Almost up until the establishment of the People's Republic, the Chinese communists enjoyed only dubious support from Moscow. The USSR continuously tried to maximize its leverage over China and pursued its interests through balancing between the Chinese Communist Party (CCP) and its nationalist rivals from Kuomintang. Only when Mao's victory in the Chinese civil war appeared certain did Stalin choose his side.[1] The Chinese leader was well aware of the Soviet dictator's dubious support of the CCP and of the ill fortune of many communist leaders whom Stalin had removed. He first traveled to Moscow in December 1949, when he felt safe as a leader of the world's most populous country.[2]

In Moscow, Mao secured a new Sino-Soviet Treaty of Friendship, Alliance and Mutual Assistance which gave the PRC much needed economic aid and military assistance. Most importantly, the Chinese got a promise of Soviet help in the case of U.S. intervention, while China's army soon benefited from a major influx of modern Soviet weapons and from training provided by Soviet military advisors.[3] The economic assistance stemming from the treaty was substantial, including a U.S.$300 million loan.[4] Yet the bargaining with Stalin was hard. The Soviet leader insisted on selling Soviet supplies to China on credit rather than giving them for free and the Chinese communists had to grant the Soviets extensive rights in Manchuria and Xinjiang and accept unfavorable terms of trade.[5]

The treaty came to a test and the two sides' relations went through further development during the Korean War. By skillful manipulation, North Korean leader Kim Il-Sung managed to get consent from both Stalin and Mao to start a conflict neither China nor the Soviet Union really wanted. Yet the three communists misperceived Washington's intentions to defend South Korea, and in a few months, the counterattacking U.N. troops under U.S. leadership were getting dangerously close to the Chinese borders. Stalin urged Chinese intervention, but fearing possible escalation, he preferred to keep Soviet involvement at a minimum.[6] Only after October 1950, when the Chinese decided to intervene, did Stalin offer substantial help in supplying his Chinese and Korean allies and moved Soviet fighters to help protect the rear of the Chinese People's Volunteers Army (PVA). Overall, the war brought the two powers closer, yet still far from full trust. A major disagreement occurred particularly about the war's termination. Stalin wanted a cease-fire in summer 1951, yet Mao, who had invested much capital into the war, disagreed at that time. But when the front stabilized along the 38th parallel, only to drain the blood and resources of the belligerent in inconclusive battles with little risk of escalation, the Soviet leader prevented termination of the conflict to use it as leverage vis-à-vis the Americans.[7]

The death of Stalin in March 1953 not only opened room for a quick conclusion of the Korean War, but also for a period of closer cooperation between the two communist great powers. Khrushchev considered China a valuable asset in the global competition between the two blocs and also a valuable asset in his domestic competition over influence with other Soviet leaders.[8] Thus, the new Soviet leadership agreed to transfer the military base in Lushun (Port Arthur) together with its equipment to the PRC, to improve the unfavorable conditions of trade, offer new loans, and after initial reluctance, start to assist the PRC with its nuclear program. In return, Beijing supported Moscow's important political steps such as the removal of Beria, the formation of the Warsaw Pact, and the establishment of relations with West Germany.[9] In this initial period, Soviet nuclear assistance was indispensable for the Chinese. In 1958, a small experimental heavy-water reactor and a cyclotron were constructed with Soviet help.[10] A number of Chinese scientists went to study in Russia while Soviet advisers arrived to assist directly in China. Shen and Xia observe that "these specialists played an important role in the selection of plant location and design and in installing and adjusting equipment."[11] At the time, Khrushchev was apparently willing to give Beijing most of the state-of-the-art sophisticated Soviet technology, including the latest designs that were not even implemented in the Soviet Union.[12] The Chinese received the Soviet P-2 short-range missiles and other modern military technology and, in an incomprehensible gesture of goodwill, Khrushchev provided Beijing with the list of KGB agents in the PRC.[13]

Despite ongoing fruitful cooperation that was scheduled to continue for some years, a grave potential for disagreement existed between Chinese and Soviet communists. The seed was sown in February 1956, during the twentieth Communist Party of Soviet Union (CPSU) congress. Khrushchev's "secret speech" denounced the brutalities and mistakes of Stalin's regime. CCP leaders thoroughly discussed the speech during several meetings in March and April. Mao, who viewed himself as a more experienced and senior leader, certainly disliked the fact that Khrushchev did not consult him in advance, yet he at least equally disliked the way Stalin treated China and Mao as an unequal partner.[14] Thus his first reaction was not truly negative, though he criticized the lack of consultations with him, felt that the criticism of Stalin should have been more specific (particularly with respect to perceived missteps in his policy toward China), and was afraid that socialism would be undermined by the leak of the speech to the West.[15] These fears were, at least in Mao's eyes, quickly confirmed by the upheavals in Poland and Hungary, which lead him to quickly change his view of destalinization from ambivalent to negative.[16]

Khrushchev's speech prepared the ideological ground for further disagreement, which was to come in two years. Dissatisfied with the development of China's economy and unable to see alternatives, Mao had decided to start a new radical economic policy, the Great Leap Forward, which in its design largely resurrected the failed ideas of Stalin's radical economic policies from the late 1920s and early 1930s. Soviet leaders, familiar with the unfortunate achievements of such policies in their own country, could hardly agree but initially kept a low profile with their objections, only gently warning their Chinese comrades about the dangers associated with their economic decisions. However, Mao refused to give himself a low profile. To Khrushchev's anger, Mao challenged the Soviets by asserting that China would accomplish communism before the USSR, while he reinforced his cult of personality at home.[17] The radicalization of China's policy also translated into a disagreement over military matters in summer 1958. The Soviets, eager to compensate for the American nuclear superiority, proposed cooperation in the realm of submarine warfare. As the first step, a joint transmitting station in China would facilitate the deployment of Soviet nuclear submarines in the Pacific. According to Jian, Soviet Ambassador Yudin "speaking on behalf of Khrushchev" also proposed to Mao the establishment of a joint submarine flotilla.[18] To Moscow's great surprise, the reaction of Beijing was furious. Mao accused the Soviets "of trying to control all of our coastline" and threatened "to wage a war against the Soviet occupiers."[19] Khrushchev, convinced that the issue was largely a matter of misunderstanding, quickly traveled to China, but achieved little, apart from humiliation when he, a nonswimmer, was made discuss matters of international politics in a swimming pool.

To make the disagreement even worse, Mao decided to rally public support for his radical policies and provoke escalation in the Taiwan Strait to attract U.S. attention to Taiwan, which to him appeared fading as the ambassadorial talks died out. On August 23, PLA artillery opened massive shelling of the nationalist-controlled Jinmen Island, a move that openly challenged Khrushchev's policy of peaceful coexistence. Mao refused to consult the Soviets in advance, despite the obligation to do so stemming from the Sino-Soviet alliance treaty and despite having met Khrushchev personally during the swimming pool meeting less than three weeks before.[20] Moscow was left largely uniformed while the second Taiwan Strait crisis escalated. As it loomed, the United States quickly reinforced their naval forces in the region, and when it turned out that the Taiwanese could not resupply the Jinmen garrison on their own, U.S. Navy started to escort Taiwanese supply ships. To the great concern of Moscow, the Americans also issued ambiguous nuclear threats and discussed the option of using nuclear weapons against the Chinese airfield in Fujian province.[21] Moscow dispatched Foreign Minister Gromyko to Beijing to find out about Chinese intentions, and publicly warned the United States to refrain from aggression against the PRC, but under the surface it was gravely disturbed by irresponsible Chinese behavior.[22]

The disagreements over the submarine issue and China's behavior before and during the second Taiwan Strait crisis made Moscow thoroughly rethink its support of the Chinese nuclear program. Khrushchev personally was greatly disturbed by Mao's repeated rhetorical suggestions that "the atomic bomb itself was a paper tiger."[23] The concerns soon transferred into the reality of nuclear cooperation. First the Soviets started to procrastinate over their aid to China's nuclear program; then they summoned their scientists back to the USSR "for vacation"; and finally, in June 1959, they informed their Chinese counterparts that Moscow would not deliver the promised

sample atomic bomb due to "ongoing international negotiations on the limitations of nuclear weapons."[24] Embittered by the Soviets' 1959 decision to
effectively terminate support for the PRC's nuclear program, neutrality in
border clashes between China and India, and negotiations with Washington
on nuclear arms control issues, Beijing retaliated a year later.

On April 16, 1960, the Chinese journal *Honggi* published an article enti-
tled "Long Live Leninism," opening what would become known as the Lenin
Polemics. For the first time, an open manifestation of the Sino-Soviet split
was made public. Prepared under the supervision of Mao himself, the article
praised Lenin's anti-imperialist policy, defending Chinese economic views
and criticizing the policy of peaceful coexistence.[25] In the next two months,
the Chinese continued their ideological offensive, first during a World Fed-
eration of Trade Unions meeting in Beijing and then at a Romanian Workers'
Party congress in Bucharest. At that time, the Soviets were ready to respond.
Khrushchev personally led the Soviet delegation to Bucharest to reply to the
Lenin Polemics, arguing that "one cannot mechanically repeat what Lenin
said many decades ago."[26] The ideological clash between the two communist
powers could hardly be more visible to fellow comrades from various frater-
nal parties, who overwhelmingly sided with Moscow. Shortly after the
Bucharest congress, Moscow recalled all Soviet civilian and military specialists from China back to the Soviet Union. The controversial decision was
partially a follow-up to the most recent disagreement and partially a response
to Chinese attempts to indoctrinate the Soviet specialists against their own
government. But while the withdrawal left China without certain economic
assistance, it also allowed the leadership in Beijing to blame Moscow for the
failure of the Great Leap Forward and left the Soviets without an important
source of information from inside of China.[27]

The following years of Sino-Soviet relations are best described as a dete-
riorating semi-conflict with some residual moments of surviving cooperation.
The alliance was virtually over, though the two sides were probably not yet
fully acknowledging the extent of conflict, particularly in 1961 and 1962
when other Chinese leaders reduced Mao's influence over Beijing's policies.
Thus in military matters, Moscow continued some support to Beijing and
even offered to deliver plans for its most recent MIG-21 fighter in 1962,[28]
and in the non-military realm, the Soviet Union provided emergency grain
supplies to help fight the famine resulting from the failure of the Great Leap
Forward.[29] Importantly, Moscow also refused to consent to Kennedy's idea
of joint anti-Chinese action.[30] But, in mid-1962, Mao's resurgence in Chinese
politics and renewed criticism of Soviet revisionism interrupted the hopes
for improved relations. Two parallel international events gave a great
impetus to Chinese criticism. First, Khrushchev was criticized for his hand-
ling of the Cuban Missile Crisis, both for the "adventurous" missile deployment to Cuba and for the "capitulation to imperialists" when the weapons

were withdrawn. Second, China convincingly won a renewed border war with India despite the lack of Soviet support.[31] Together the two crises increased Mao's confidence in China and his conviction that the Soviet Union was in decline.

As the relations between the communist great powers worsened, various old issues along the poorly demarcated 4,500 miles long border resurfaced.[32] During a press conference in Katmandu on April 28, 1960, Zhou Enlai gave subtle hints that "some disagreements existed between Chinese and Soviet maps."[33] In two months, China started using the strategy of sending its cattle herders to the Soviet territory to settle there and stay despite demands by Soviet border guards to leave. Incidents of such kind continued over the course of following years.[34] A different major incident occurred in Xinxiang where a small-scale rebellion against Chinese authorities started. Though the details are blurred, it is clear that tens of thousands people, many with Soviet passports, left China for Soviet Kazakhstan.[35] Starting from 1962, both governments began reinforcing border garrisons in the respective regions and, in March 1963, the Chinese government-run *People's Daily* publicly referred to an unequal nature of the treaties that demarked the Sino-Soviet borders.[36]

Despite initial hopes on both sides and Mao's personal satisfaction with the fall of his enemy, the removal of Khrushchev in October 1964 changed little in Soviet policy toward China. The new leadership in Moscow was divided over the prospects of relations with the PRC. Premier Aleksei Kosygin was the most eager supporter of rapprochement, while Secretary General Brezhnev did not take any position and others like Yurii Andropov were skeptical. But none of the Soviet leaders was ready to cede leadership of the communist bloc to Mao, probably the only thing that could have really repaired mutual relations.[37] This appeared in the clearest possible sense when Mao dispatched Zhou to Moscow to meet with the new Soviet leaders. After Zhou's arrival, Brezhnev made it clear that the Soviets insisted on peaceful coexistence with the West and upheld the year-old test ban treaty. To make matters worse, the drunken Soviet Defense Minister Malinovsky suggested to Marshal He Long that the Chinese should follow the Soviet example and "get rid of Mao Zedong." The Chinese protested, and having received radical instructions from Beijing, they refused to accept a Soviet apology for Malinovsky's behavior. Then, through the words of Anastas Mikoyan, the angry Soviets told the Chinese delegation that "there was no difference of opinion between the new leadership and Khrushchev on the question of the basic causes of the Sino-Soviet dispute."[38] Soon the only major change of the Soviet course with respect to China was the abrogation of Khrushchev's policy not to support Vietnam until it firmly chose its side in the Sino-Soviet dispute.[39]

Sometime between 1964 and 1966, Mao came to the conclusion that the Soviet Union would attack China.[40] This idea had the potential of a self-fulfilling prophecy. As Joseph argues, Mao was gravely afraid that "without decisive remedial action, socialist revolutions inevitably degenerate in revisionism," but in fact he could do little beyond his polemics to challenge Soviet revisionism.[41] To prevent such degeneration at home, the Chinese leadership started the unfortunate Cultural Revolution in 1966, further radicalizing Chinese politics, society, and foreign policy, and bringing Beijing into self-imposed international isolation with hostile powers all around its borders. This was unwelcome news for the Soviets, who had previously on numerous occasions concluded that they did not understand what the – in their eyes unpredictable – Chinese wanted.

Mao's fears greatly worsened in August 1968 when almost 500,000 Warsaw Pact forces occupied Czechoslovakia, effectively ending the liberalization processes of the Prague Spring and then in November when Brezhnev explained the rationale for the intervention by stating that the threat of restoration of capitalism in one socialist country is "the concern of all socialist countries."[42] Afraid that its allies in Eastern Europe or China itself would become the next target, Beijing condemned the intervention, starting to use a new label of social-imperialism to describe Soviet behavior.[43] By that time, incidents were occurring regularly on the Sino-Soviet border, though casualties were usually avoided. Both sides blamed the other. The regular incidents made Beijing and Moscow reinforce their military in the contested regions, and as a result, significant forces gathered on both sides of the border. By 1968, 16 Soviet divisions with heavy weapons along the Sino-Soviet border and six in Mongolia faced 47 lightly armed Chinese divisions.[44]

The conflict abruptly intensified in March 1969. The first battle on the small island of Zhenbao (Damanski) opened on March 2, through what current historians consider to be a Chinese ambush.[45] The actual course of events is less clear. Most likely, the Chinese stationed around 300 selected PLA elite troops under the cover of darkness during the night from March 1 to March 2 and then dispatched some 30 men during the day to attract the Soviets' reaction. When Soviet border guards arrived and, according to a procedure established from previous incidents, ordered the Chinese to leave, the shooting started. The Soviets were taken by surprise. After inflicting heavy casualties on the Soviets, the Chinese forces probably withdrew, claiming a clear victory, while Soviet reinforcements took control of the island.[46] The second, larger, clash occurred on March 15. By that time both sides had made preparations for the clash and moved heavy weapons into the vicinity of the island. The initiator cannot be recognized; most likely the Soviets dispatched troops to take control of the island, but met stronger Chinese resistance than expected. Soviet communications were failing; the border guard had difficulty to secure help from the regular army and its commander Colonel Leonov was killed in action. Only after communication was established with Brezhnev, at the time visiting Hungary, did the regular Soviet army join the battle. The Chinese were then repelled, but they prevented the Soviets from holding the island with artillery fire.[47]

It appears that the Chinese had a twofold rationale for the escalation: first, to mobilize domestic public against the Soviet menace and thus to end the most intensive phase of the Cultural Revolution;[48] and, second, Mao wanted to teach the Soviets a "bitter lesson" and thus deter further border violence.[49] However, the immediate effects were completely different. Old fears over the vulnerability of the Soviet Far East resurfaced as the Chinese caught the USSR unprepared and the Soviet leadership felt that some radical action might be necessary to deal with the Chinese threat.[50]

Without access to archives in Moscow, full details of the decision making of the Soviet leadership are as yet impossible to reconstruct. Certainly the clashes on Zhenbao were unprecedented; open hostilities of such kind never happened between the Soviets and the Americans. In this situation, radicals lead by Defense Minister Andrei Grechko reportedly advocated that the Soviet answer should be a major nuclear strike. The army leadership was more cautious: Deputy Chief of Staff Nikolai Ogarkov suggested that a limited nuclear attack aimed at the Chinese nuclear arsenal and related facilities would be more appropriate, should nuclear escalation be required.[51] On the other side, Prime Minister Alexey Kosygin led the peace party and tried to negotiate with Beijing since the beginning of the crisis. The position of Brezhnev cannot be determined given the lack of available evidence, which may well correspond with his standard operational procedure not to take a clear stance on controversial issues during the Politburo's foreign policy discussions.[52]

Kosygin, who was the most optimistic about the possibility of rapprochement with Beijing after the fall of Khrushchev, took the lead in the attempts to calm the situation. On March 21, he tried to reach Mao on the hot line, but shockingly the operator called the Soviet premier a "revisionist element" and refused to connect him with Mao.[53] On March 29, Kosygin publicly issued a conciliatory statement calling for negotiations on border issues.[54] But this time, the Soviet premier accompanied his plea with a threat, warning the Chinese that Moscow was ready to counterattack should the provocations continue.[55]

The clashes continued nonetheless, though on a smaller scale, during spring and summer. The most serious incident appeared in Xianjing on August 13. This time the Soviets were most likely the aggressors, choosing a battlefield remote from Chinese bases but close to the Soviet railhead. The Soviets reportedly encircled the Chinese troops with tanks, killing some 20 soldiers in a region where China felt extremely vulnerable.[56] In the meantime, Moscow was reinforcing Soviet troops at the border regions, but with only the Tran-Siberian railway available as a suitable supply line, it would take at least till 1972 to match the Chinese forces with 18–20 additional divisions being deployed to the Russian Far East.[57]

According to Gobarev, the continuing incidents made the Soviet leadership believe that Mao "would come to the negotiation table only after he realized that a Soviet nuclear strike against Chinese nuclear installations was imminent," thus in mid-August they decided to threaten a nuclear strike.[58] At this point Gobarev may exaggerate; it is unclear whether the Soviets really contemplated striking China with nuclear weapons. Likely Defense Minister Grechko was in favor, but Gobarev himself argues that Brezhnev and other senior leaders "were committed to the principle of no first use."[59] Recent researches by Goldstein also suggest that a conventional strike was the preferred option of the Soviet leaders.[60] To make such an option possible, Soviet bomber units were transferred from western Soviet Union to the Far East and, according to Lewis, "practiced bombing runs in preparation for a strike on Chinese nuclear facilities."[61]

Soviet style signaling started early in the crisis and suggests that Moscow did not take the crisis lightly. Already on March 8, the Soviet Army newspaper *Krasnaya Zvedza* printed the first nuclear threat of the crisis. In August, the same paper published an article by the new commander of Soviet forces in the Far East, General Tolubko, who commemorated the successes of the Red Army in repulsing Chinese incursions in the 1920s.[62] With the fighting in Xianjing and Tolubko's implicit threats, August became one of the hottest moments of the crisis. Either to secure American consent, or to increase the pressure on Beijing, Boris Davydov, officially the second secretary of the Soviet Embassy in Washington and actually a KGB officer, questioned William Stearman, a midlevel State Department official, about possible U.S. reaction to the Soviet strike against China's nuclear installations on August 18.[63] Reportedly, at roughly the same dates, similar informal inquiries occurred with the Soviets' European allies. On August 28, an editorial in *Pravda*, a major Soviet newspaper, warned the world of how dangerous China was.[64] By no means did Beijing see the Soviet threats as empty-worded; on the same day when *Pravda*'s editorial was published, the CCP Central Committee issued an "Order for General Mobilization in Border Provinces and Regions."[65]

An opportunity to de-escalate came abruptly on September 2, when respected Vietnamese leader Ho Chi Minh died. Since the time of Brezhnev's leadership decision to overturn Khrushchev's policy of not supporting Vietnam until it chose sides in the Sino-Soviet dispute, both China and the USSR maintained friendly relations with Vietnam. Moscow and Beijing thus dispatched high-ranking delegations lead by prime ministers to attend Ho's funeral. In Hanoi, Zhou and Kosygin briefly spoke about the issue, but Zhou was evasive arguing that the he first had to consult Mao.[66] Kosygin stayed in Hanoi for a few days waiting for a Chinese response but then, after receiving no message from his counterpart, he decided to return to Moscow. However, while the prime minister was en route the Chinese approached Soviet *chargé d'affaires* Elizavetin requesting negotiations with Kosygin. Reached by the Chinese message in Tashkent, Kosygin turned his flight to Beijing and arrived on September 11.

The September 11 meeting between Kosygin and Zhou Enlai at Beijing Airport appears to be the breaking moment of the crisis. From the beginning, the possibility of Soviet preventive destruction of China's nuclear installations was discussed. Zhou apparently devoted a considerable effort to convincing the Soviets that the PRC and its nuclear program were no threat to the USSR and that China did not have territorial claims toward the Soviet Union, despite the fact that it considered the treaties that delimited mutual borders unfair. In return, Kosygin attributed the alleged preparations of a Soviet preventive strike to imperialist propaganda. According to the Soviet account, both prime ministers also agreed on the principles for de-escalation, most importantly observation of existing borders, inadmissibility of armed confrontations, and withdrawal of troops from direct contact in controversial sectors.[67] However, Zhou did not limit himself to conciliations. He not only insisted on a continuation of ideological polemics, but he also threatened that if the Soviets "take preemptive measures to destroy our [Chinese] nuclear facilities … [PRC] will fight to the end."[68] According to Goldstein's recent research, through the aforementioned words Zhou threatened the Soviets with a long-lasting guerrilla war not dissimilar to the one the Chinese had waged against Japan.[69]

With the advantage of hindsight, but with archival sources in short supply, the Kosygin–Zhou meeting seems to be crucial for defusing the crisis. Though the leaks about the possibility of a Soviet strike appeared again on September 16, it is likely that Kosygin's "peace side" prevailed against Grechko and the hawks. The Soviets gradually stopped the preparations for airstrikes in the following years and arguably decided to counter the Chinese threat to the Far East with a major buildup of conventional forces in the region and by a formidable yet incredibly expensive boost of the Soviet Union's strategic position in the region by building a northern alternative to the Trans-Siberian railway.[70] It took at least several months for Beijing to accept that the Soviet strike was not imminent. In fact, the Chinese initially considered the Zhou–Kosygin negotiations rather to be a cover-up to mask Soviet intentions. Anticipating that the Soviet attack would come on China's National Day, Lin Biao ordered the PLA to be on full alert on September 30, and when border negotiations were scheduled to start in Beijing in October, the Chinese top leadership was dispersed to survive a Soviet attack on Beijing and to command the national resistance against Soviet invasion.[71] Only gradually, over the course of the next months, did the Chinese leaders accept the fact that the Soviets had decided not to attack.

Unfolding the complexity

Saying that, similarly to the previous case, the Sino-Soviet border war and related contemplation of a preventive strike in Moscow show the complexity of deterrence relations is probably unnecessary. So how does the conundrum of the Sino-Soviet case appear if split into the conceptual framework of this study?

By the end of the crisis, China's nuclear arsenal was almost five years old, but hardly beyond its infancy. China first tested its atomic bomb in 1964 and its airdrop in 1965.[72] Through the sheer numbers, the arsenal had almost passed the period of smallness. Kristensen and Norris, whose dataset is highly respected, estimate the number of Chinese atomic weapons at 50.[73] However, other necessary parts of the nuclear complex were far less advanced. One problem was the availability of reliable delivery vehicles. In 1966, China tested the DF-2 missile with a 20 kt warhead and a range of 1,455 km and the DF-3 with a range of 2,800 km. Both liquid-fueled weapons were vulnerable to pre-emption and offered only limited accuracy in the absence of a proper inertial guidance system.[74] Also, while information about entry into service is not entirely clear, most sources attribute the missiles' deployment to combat units as late as 1971–1972.[75] As an alternative to missiles, some 150 Soviet-designed Il-28 jet bombers were available to China, even though it is unclear how many of them were nuclear-capable. Finally, considering the fact that China only launched its first satellite in 1970, it is more than likely that its early warning system was underdeveloped and, as Goldstein argues, its "command and control capabilities were primitive at best."[76]

This was in stark contrast to the 1969 Soviet force, which at the time arguably entered the decade of its greatest might. After Khrushchev's missile gap bluff in fact opened an even greater real gap between the feigning Soviet strength and the real American one, the Soviet Union embarked on a fast track of catching up with the Americans, introducing new weapons on an unprecedented scale. By the end of 1970, the Soviets deployed 1,290 ICBMs and 300 SLBMs.[77] Almost all of the ICBM force comprised of SS-11 liquid fuel, single warhead missiles with one-megaton warheads and a fairly low accuracy at around 1.4 km CEP.[78] The Soviet bomber force, arguably the most appropriate tool for a surgical preventive strike at the moment, also underwent substantial improvements. By 1969 the Soviet Union possessed substantial numbers of modern subsonic Tu-16 (Badger) and Tu-22 (Blinder) supersonic jet bombers, some of them capable of equipping air-launched cruise missiles.[79] Far less developed was the Soviet early-warning system, which only deployed fully operational dedicated radar systems to protect Moscow in 1970 and launched the first generation of early-warning satellites in 1972–1979, but which already utilized the airborne early-warning and control Tu-126 aircraft.[80] Despite deficiencies in some areas of the Soviet nuclear complex, the asymmetry between the two countries was enormous, likely with profound implications for China's ability to withstand a Soviet attack.

The Chinese appeared to be acquainted with the limits of their nuclear deterrence. Most interestingly, Mao noted in a conversation with Australian communist E.F. Hill that "if we are to fight a war, we must use conventional weapons."[81] Similarly, Zhou conceded to Kosygin at the airport meeting that China was not able to fight a nuclear war.[82] The doubts were not unwarranted. The notable Soviet shortcomings in early warning and possibly in their command and control system would not influence their ability to strike first, but rather enhance their motivation to do so. While China already had a number of atomic bombs available, increasing the chance that some would survive the Soviet attack, Beijing lacked the appropriate means to deliver the counterblow. Not only is it uncertain that any available Chinese missiles were operational and armed with nuclear warheads at the time of the crisis, but even if they were, they would be suitable targets for Soviet bombers due to the long time required for launch preparation. The substantial fleet of Chinese Il-28s would have likely warranted survivability of at least some aircraft, but this old 1950 vintage bomber could reach valuable targets in the European USSR only under favorable conditions on a one-way mission; in fact, they were hardly able to penetrate through Soviet air defense under any conditions even against targets in the Asian part of the USSR. The location of critical Chinese nuclear installations was also not well chosen for a war with the Soviets. Most of them were constructed with the help of the Soviet Union and located precariously close to the Soviet borders, casting further doubts on the Chinese ability to meet the second strike criterion. In fact, virtually all experts express doubts about the survivability of the Chinese nuclear arsenal vis-à-vis a possible Soviet preventive strike, and their view is consistent with Mao's statements from the time.[83]

The conventional balance was more complicated. In 1969–1970, the Soviet Union deployed 147 divisions, but most of them were designated for a central-front battle against NATO and stationed in Eastern Europe or the European USSR. Despite China's advantage in manpower, the PLA had only 115 divisions, but almost half of them at the Sino-soviet border.[84] After the Sino-Soviet split the PLA lost its most valuable source of modern arms and only slowly did China start to produce copies of Soviet weapons and to improve them. At the same time, Mao's experimental Cultural Revolution did little to improve the combat readiness of the PLA, particularly as it decimated its officer corps.[85] The Soviet advantage in technology is beyond any doubt. The situation of the tank force is illustrative. In the battle on Zhenbao, the Soviets already deployed the new T-62 tanks, which were already available in substantial numbers, while the even newer T-64 tanks had just started entering the service. The PLA, on the other hand, had to rely on type-59

MBTs, which in fact were a Chinese version of the successful but already 20 years' old Soviet T-54 design and which, according to eyewitness, had "many problems, including weak armor protection, poor mobility, and lack of communication."[86] In fact, the PLA did not even use tanks at Zhenbao and after the clashes the Chinese celebrated a victory of their spirit against the technologically superior yet morally inferior Soviet troops.[87] It is unlikely that this alleged moral superiority would have stopped a full-scale Soviet invasion. By 1970 the Soviets had for the first time developed the means to project conventional power far beyond their borders,[88] and were enjoying substantial general conventional preponderance. The Chinese attempted to meet that challenge by asymmetric warfare, which they had mastered in a war against Japan and which they were planning to use during the October panic when the PRC leadership was dispersed not only to survive the expected Soviet attack, but also to lead local guerillas against the invasion.[89]

Contrary to the global military situation which clearly favored the Soviets, circumstances in the border regions were less favorable for Moscow. The Soviets were, of course, able to transfer their technological superiority into the battlefield, as was the case in the first combat deployment of GRAD surface-to-surface multiple rocket launchers in Zhenbao.[90] However, the Chinese numerical superiority was a great offset. By 1968, the USSR forces in the border regions with China were limited to 22 divisions, 16 on the Soviet soil and six in Mongolia.[91] Reportedly at least some of those units were under strength.[92] The Chinese manned the borders with 47 lightly armed divisions, enjoying an advantage in manpower of roughly 2.5 to 1.[93] With the exception of Xinjiang, the PLA also benefited from shorter lines of supply, compared to the Soviets who had to rely on the Trans-Siberian railway which runs precariously close to the Chinese border. None of the shortcomings was secret to the Soviets. The crisis triggered a major redeployment of Soviet forces, with 18–20 divisions moving against China in 1969–1972 period, but most of them were not available in summer 1969.[94] The vulnerability of the Trans-Siberian railway also made the Soviets build the costly and otherwise fairly useless Baikal–Amur Mainline, which could serve as an alternative supply line for Russian Far East forces.

As it has been noted several times, the Soviet technological advantage was beyond any doubt. The aforementioned tank example was rather a rule than an exception. China completely lacked some major weapons systems, including nuclear-powered submarines or short-range artillery rockets such as GRADs and FROGs. Its bomber force was largely obsolete and the Cultural Revolution prevented China from fully utilizing its technological potential even where it could. Thus while China managed to produce Soviet MIG-21 fighters whose design it got from Khrushchev in 1962 and which were standard Soviet fighters (though in modernized versions) at the time of the border clashes, the production proceeded very slowly well into the

1970s.[95] The PRC was also lagging behind in the utilization of space technologies. Its first satellite was only launched in 1970, while the Soviets had already deployed their Zenit 2 photoreconnaissance satellites.

The Zenit intelligence was not the only source of Soviet information about the Chinese nuclear program. To improve available overhead imagery and compensate for the deficiencies of the still relatively unsophisticated Soviet satellite intelligence, the USSR conducted a large number of overflights over China's nuclear installations.[96] According to the memories of Soviet witnesses and the accounts of Russian experts that have been gathered by Goldstein, an even greater role was played by Soviet intelligence services. Former Soviet *chargé d'affaires* Elizavetin recalls that: "the Soviet military leadership was at this time extremely well acquainted with the capabilities of Chinese armed forces, because ... they were built with their indispensable participation." Soviet general Anatoly Boliatko also agrees, arguing that "Soviet intelligence possessed the targeting information required to carry out a disarming first strike."[97] Without the ability to access Soviet archives and compare what was available to the Soviets with actual Chinese deployment, which is unlikely to happen anytime soon, the accuracy of those accounts cannot be verified. However, the ability of Soviet intelligence services to penetrate various Chinese organizations is not unlikely. Before the rift between Soviet and Chinese leaders terminated the cooperation, a large number of Chinese civilian experts and military officers went to study in the Soviet Union, while a large number of Soviet advisors worked in China. Soviet intelligence must have been in touch with this large pool of people and certainly established working contacts.

The Soviet Union had a military edge over China in nuclear and, with the exception of the border regions, in conventional weapons, it employed more advanced technology, and it had confidence in the accuracy of targeting information. However, military factors are only one part of the story and must be combined with political perceptions into a complex picture.

Certainly the political and military conflict between the Soviet Union and China stood high among the priority issues of leaders in both Moscow and Beijing. In relative terms, the Soviet threat clearly became the greatest priority of the Chinese leaders. With the heating up of the Sino-Soviet split, the Soviet Union surpassed the United States as the gravest threat to the PRC, earning the title of a social-imperialist country in Chinese ideological discourse.[98] However, for the Soviet leaders, the situation was more complex. Certainly the ideological conflict with China was central to the Kremlin. Yet the USSR was a global superpower with worldwide interests. The global competition with the West was at least equally important for the Soviets, who would have often preferred not to be disturbed by the China issue.[99] However, it also appears that at least for a brief period, Zhenbao may have changed the situation. Open shooting which left tens of the Soviet Union's soldiers dead was unprecedented in its relations with the Americans.[100] The United States was certainly perceived as a more formidable adversary, but with détente looming, the centrality of conflict between the U.S. and the USSR declined, or more accurately the centrality of the relations between Moscow and Washington remained unquestionable in the eyes of the Kremlin's rulers while the conflicting part was reduced. While it remains debatable whether the conflict with China was able to occupy the most central position in relative terms, its centrality in absolute terms is as clear as the unpleasantness of the situation on the borders for the Soviet government.

To the Kremlin's leaders, the Chinese were an unpleasant enemy, one which appeared rather unpredictable, irrational and, to make matters even worse, very resolute. The contrasting Soviet perceptions of the Americans, their main Cold War adversaries, as predictable and rational – and of the Chinese as irrational and unpredictable – bears a surprising similarity to Kennedy's views of the Chinese and the Soviets. It would be a fairly lazy approach to copy and paste the appropriate paragraph from the previous chapter, but in a sense it would be materially correct; the same factors that shaped the U.S. view of Chinese resolve largely applied to the Soviets as well. The Chinese stood firm against the U.S. nuclear threats, accepted heavy casualties in Korea, and had experience of a war of resistance against Japan. To remind Kosygin of the Chinese ability to fight long wars, Zhou explicitly threatened him during their airport meeting that "his nation would consider itself to be in a war to the end with the USSR," trying to impress his Soviet counterpart and persuade him about Chinese resolve.[101]

Yet Kosygin's meeting with Zhou also signifies a factor that is markedly different from the previous case. Contrary to the Americans who, apart from unofficial ambassadorial talks, interrupted their communication with Beijing for more than 20 years, the Soviets kept the communication channels open. The USSR downgraded its representation in Beijing to the *chargé d'affaires* level, yet the practical difference it made from the deterrence perspective was small; a diplomatic channel was available. A direct phone line apparently also existed between the leaders in Moscow and Beijing, though Kosygin's March 21 unsuccessful attempt to reach Mao, which only earned him the pleasure to be called a "revisionist element" by the Chinese operator, put its utility into question.[102] To what extent the existence of communications channels and at least 10 years of recent experiences with close relations between Beijing and Moscow created a set of shared expectations about the requirements of stable deterrence is rather uncertain. The Soviet leaders repeatedly felt that they did not understand what the Chinese wanted and such misunderstandings prevailed even during the golden years of the Sino-Soviet alliance. This makes it difficult to judge the institutionalization of mutual relations in terms of limited or strong. Sufficient communication channels existed, but the transfer of information was blurred by differing expectations and understandings on both sides. However, I argue that despite the deficiencies, the dyad can be treated as strongly institutionalized, considering the close working experience of many Chinese and Russians that were made in the 1950s and the existence of communication channels.

This goes hand-in-hand with a long but limited history of hostility between the two countries. The relations between China and Russia date back several centuries. Parts of mutual borders were first demarcated by the treaty of Nerchinsk in 1689, with other treaties following over the course of a century. During the mid-nineteenth century, tsarist Russia took advantage of China's weakness and acquired large portions of mostly uninhabited territory that the previous treaties assigned to China, and made Beijing consent to the changes in several treaties.[103] Later Chinese leaders, including Mao, considered the treaties unequal and occasionally raised the issue with the Soviets, even during the most cooperative times. Yet the border regions had in fact little value and the border topic was mostly raised by the Chinese to support other disputed issues.[104] More serious in the eyes of the Chinese communists in general, and Mao in particular, was the way Soviet leaders treated their Chinese counterparts as junior partners and later ideological disagreements. Yet this hostility was hardly anchored in history. In fact, despite occasional setbacks, the period from 1949 to at least 1959 is well described as cooperative, however careful this cooperation sometimes was, and with full extent of the conflict not coming for at least several years after 1959.

While the history of hostility between the People's Republic of China and the Soviet Union was limited, another Asian experience – the 1905 Russo-Japanese war – may have influenced the perception of some Soviet leaders. The lesson of a humiliating Russian defeat certainly amplified the fears and concerns over the defensibility of the Soviet Far East against possible invasion. However, it does not appear that last-resort logic was present either in the Soviet hawks' or in the Soviet doves' thinking. The available evidence suggests that even Grechko and likeminded officials advocated "unrestricted use of multi-megaton numbs" against Chinese targets to "once and for all get rid of the Chinese threat" – rather on the basis of "better now than later" than as a last resort.[105] Certainly the situation did not appear as the last resort to the decisive majority of Soviet leaders, who apparently contemplated other alternatives to striking China's nuclear installations and opted for a conventional buildup at the border.

This absence of "last-resortness" in Soviet thinking appears surprising, considering the Soviet perception of Chinese leaders as highly irrational. The roots of this view go back to the 1950s, in particular to the Taiwan Strait Crisis. Mao repeatedly expressed his disrespect for American nuclear threats. In late 1954, Mao told Indian Prime Minister Jawaharlal Nehru that the more populous socialist camp would survive a nuclear war while "the imperialists would be totally wiped off the face of the earth."[106] Three years later, Mao stunned the delegates with the same ideas at a meeting of communist parties in Moscow. In front of Khrushchev and other top leaders of communist countries, Mao expressed his belief that perhaps one half of the world's population would be destroyed in a nuclear war, but that this war would lead to a complete destruction of the imperialists followed by the dominance of socialism and recovery after the couple of years. Other delegations were shocked. Czechs and Poles objected immediately.[107] But things appeared to go from bad to worse. A year later, Mao started the Taiwan Strait crisis. When Foreign Minister Gromyko arrived in Beijing to find out about Chinese intentions he learned from Zhou that "PRC is now ready to accept on itself all serious blows right up to the atom bomb and to the destruction of our cities" and that the USSR was then supposed to respond with a nuclear counterstrike on the U.S.[108] This was enough for the Soviet leaders and added substantially to their decision to withdrew Soviet help from China's nuclear program. The end of cooperation may have had an unwelcome negative side effect. Over the course of years, Mao changed his understanding of nuclear weapons, coming a long way from his "paper tiger" concept.[109] Yet the split probably prevented the Soviets from identifying the change. On the contrary, Mao started the Great Cultural Revolution and drove China into chaos, further contributing to the Soviet perception that he was an irresponsible and irrational leader.

Unfolding turbulent international events in the background of the Sino-Soviet clashes certainly deserve equal attention to perceptional issues. The United States policy had a prominent role there as it was the only country in a position to provide meaningful third-party deterrence vis-à-vis the Soviet Union. It is also well known that by 1969 the Nixon administration had started the process of rapprochement with China.[110] The declassified American evidence also shows that Washington was gravely concerned about the possibility of a Sino-Soviet war. President Nixon explained to his cabinet on August 14 that "the worst thing that could happen for us would be for the Soviet Union to gobble up Red China." Reportedly, the Americans shared their disapproval with the Soviets. According to Betts, Soviet Ambassador Dobrynin reported to Moscow that "the U.S. would not be passive regarding such a blow at China."[111] However it is uncertain how strongly Moscow's decision making was influenced by the U.S. position. There, Goldstein's arguments are persuasive. The U.S. public strongly disapproved of the one war that their country was already involved in and it is hard to believe that it would be more sympathetic to defending communist China. Furthermore, the United States did not have the conventional capacities to threaten the Soviets, while their nuclear threats would hardly have credibility vis-à-vis the threat of Soviet nuclear retaliation.[112] The patterns of Soviet interventions in Hungary, Czechoslovakia, or Afghanistan show that the Kremlin carefully weighed the threat of U.S. intervention, but where it appeared safe that a war with the United States would not come, it readily accepted the political costs the U.S. could impose.

A similar analogy likely applies to Soviet sensitivity to alliance politics and to international action legitimacy, which in this case are deeply inter-linked. After the March 1969 clashes, the Soviet leaders appealed for help at a Warsaw Pact summit, asking their allies to send troops to the Soviet Far East, but the allies expectedly refused Moscow's request.[113] This was hardly a striking blow for Moscow. Materially there was little the allies could add to the Soviet power. On the other hand, the refusal to dispatch troops not-withstanding, politically most communist countries stood firmly behind the Soviets. Only Romania and North Vietnam pushed the Soviets to seek a negotiated solution, but their leverage over Moscow was limited.[114] Previous and subsequent Soviet interventions reveal that while the Soviets sought help from their allies and utilized their military participation, they did so mostly to enhance action legitimacy. Despite the allies' valuable participation in the occupation of Czechoslovakia, Moscow had demonstrated its readiness to intervene militarily on its own in Hungary, when it believed the situation required a military solution. Certainly the Soviets had little respect for the norm of nonintervention into other socialist countries' affairs, a fact they publicly articulated in the Brezhnev doctrine.[115]

The consistency of Soviet policy with respect to the nuclear taboo creates an interesting observation. The taboo literature has paid predominant atten-tion to Western, mostly American, evidence with respect to the taboo, and Tannenwald herself admits that "taboo is probably not universal," though in another place she cites Soviet willingness to rather accept defeat than to use nuclear weapons in Afghanistan as an evidence for the taboo.[116] The evid-ence from the Sino-Soviet case is supportive of the Soviet commitment to no first use of nuclear weapons. Defense Minister Grechko allegedly suggested solving the Chinese problem once and for all with a massive strike with multi-megaton nuclear weapons, while Chief of Staff Ogarkov suggested a surgical strike with a few tactical weapons. However, these options were apparently not widespread and not approved by the Politburo. According to Gobarev, a majority of Politburo and Central Committee members, though hawks, preferred conventional weapons, but only because they believed they would be sufficient, while the military opted for a combination of both con-ventional and nuclear arms, the latter to be used in the last resort.[117] The actual effect of normative factors on the Soviet decision is not clear, yet it appears that the Soviets were strongly committed to preferring conventional weapons on consequential reasons.

The diverging opinions of various leaders and factions in Moscow high-light the need to turn to domestic factors. Of them, domestic action legiti-macy appears to play only a very limited role, a view that is consistent with the expectations of lower constraints faced by nondemocratic countries in this realm. Neither literature nor available evidence suggests that domestic action legitimacy was considered prohibitive in the Kremlin, although for the Soviet population China was largely an unnatural enemy. Thus despite limited domestic legitimacy, the Soviet leaders were not constrained in their decisions about attacking the Chinese nuclear complex.

More interesting in the domestic policy realm is the opposition among influential decision makers to anti-China strikes. A split between hawks and doves apparently existed in the Soviet leadership. The China Department of the Communist Party Central Committee was the center of anti-China hawks who believed that a confrontation and perhaps war with China was inevitable, a view that was also widespread in the Ministry of Defense, in the Soviet Army General Staff, and in parts of the KGB. Yet the KGB was divided over the issue. The Information-Analytical Directorate under Major General Leonov believed that the Soviet main interest lay in Europe while the Chinese actions in Southeast Asia were making no clash necessary. China doves also dominated Soviet military intelligence who got accustomed to cooperation with China in the 1950s as well as in state (rather than party) institutions such as the Foreign Service.[118] The institutional feelings largely corresponded with the opinions of top decision makers. The most ardent supporter of an anti-China strike appeared to be Defense Minister Grechko. On the contrary, Prime Minister Kosygin who headed the state institutions was the most important actor of reconciliation. Kosygin was consistent in his views on China. Already after the departure of Khrushchev, Kosygin pressed for an improvement in the relations with the PRC.[119] During the crisis he put an enormous effort into negotiating a peaceful solution and his meeting in the Beijing airport appears to have been instrumental in this. The most difficult to uncover is the actual position of Brezhnev. As Zubok describes, the Secretary General "shared racism-colored fears of China. He neither trusted Maoist leadership nor wanted to negotiate with them, leaving the unpleasant business to Kosygin."[120] However his position during the crisis remains unclear, which is consistent with his behavior during the Politburo's foreign policy discussions where he regularly avoided taking a clear stance on controversial issues.[121]

Turning to regime type, it is clear that the 1969 Soviet Union was a highly authoritarian nondemocratic regime, scoring −7 in the widely respected Polity IV database.[122] More interesting is the extent of the military's influence over the political leadership. There, the Soviet system was often misunderstood. It was supposed that the Soviet military had a much greater influence than its Western counterparts due to instances of symbolism inconceivable in the West such as active duty officers serving as ministers of defense. However, the communist state in fact created a highly sophisticated divide-and-rule system of balancing the military's interests and protecting itself from a coup. Soviet armed forces were fragmented, with powerful parts such as KGR troops and Ministry of Internal Affairs troops standing outside the army's authority, and controlled from inside by the powerful Main Political Administration of the Communist Party and special sections of the KGB.[123] During the prewar and Second World War period, the military learned that it was safest to focus on defense issues and not to get involved in politics. While Brezhnev's leadership provided the army with abundant resources and certainly respected its professional authority and expertise over defense subjects, the army did little to challenge political judgments.[124]

Critical, at least from the perspective of deterrence theory, is how the conundrum of the 1969 Sino-Soviet war performs with respect to the four factors of military threats. Expectedly, there is little to say about nuclear denial; the Chinese were neither planning to use such strategies, nor threatening the Soviets with them and, on the other side, the Soviets were not afraid of Chinese nuclear denial.[125] However it also appears that the threat of Chinese nuclear retaliation was not a dissimilar story.

While Mao reportedly threatened that "China possessed its own nuclear bombs and could take a terrible revenge" it does not appear that China seriously considered using nuclear weapons against the Soviet Union.[126] During the heat of the crisis, Chinese leaders did not order China's nuclear forces to prepare for striking back, but to prepare for accepting a strike.[127] It seems that the contours of this situation were known to the Soviets. Even though it is impossible to fully reconstruct the Soviet top decision makers' sensitivity to the threat of Chinese nuclear retaliation without access to the archives, Russian interviewees questioned by Goldstein – whether Russian specialists in the field, or eyewitnesses in the lower ranks of the Soviet hierarchy – agree that the threat of Chinese nuclear retaliation was not deterring Soviet military action. This confidence of the Soviets was largely attributed to their ability to break the Chinese capability to retaliate in a first strike thanks to the targeting information from Soviet military intelligence.[128] Also, the behavior of the Soviets during the crisis suggests that nuclear retaliation was not of the greatest concern to Moscow. They put a lot of effort into mitigating the Chinese conventional threat, redeploying units from the European central front to the Far East in a fast pace, but did little to deal with the option of nuclear retaliation, apart from enhancing the already formidable Soviet first-strike capacities vis-à-vis the Chinese with additional bombers and nuclear missiles in the region.

The threat of China responding with an escalation of conventional war was probably of much greater concern to the Kremlin's leaders. Wiegand argues that the possibility of "having to tie down large forces for recurring, indecisive battles against an enemy specializing in hit-and-run warfare and disposing of inexhaustible manpower" was a nightmare for Soviet leaders.[129] Such a description is hardly inaccurate. The possibility of a large-scale Chinese invasion made Moscow unsure about its own prospects in such a war. Military leaders both at the Ministry of Defense and in the General Staff and the KGB were gravely afraid of China's vast manpower, which the Soviets could only match by the massive use of nuclear weapons on the battlefield.[130] The Chinese well understood their advantage in numbers and were not reluctant to use such a threat to challenge the Soviets. Thus at the critical Beijing airport meeting, Zhou did not forget to remind his Soviet counterpart that "his nation would consider itself to be in a war to the end with USSR" recalling that the conflict would not be over with the destruction of China's nuclear complex.[131] Consequently, the Soviet reaction corresponds with the sensitivity to China's conventional might. Moscow went to great lengths when redeploying forces from the Cold War's main battlefield in Europe, where traditional Russian interests lay, to the remote and sparsely populated areas at the Chinese border, and paying enormous costs for the economically unsound Baikal-Amur Mainline.

The threat of conventional denial, while more robust than the nuclear one – at least as it existed – also appears to have only a limited effect on Soviet decisions. Certainly the Chinese nuclear weapons complexes had substantial air defenses that would have made a Soviet strike costlier. Nevertheless, the available empirical evidence does not give any indications of Soviet fears of a Chinese ability to prevent the Soviets from destroying the targets or to make their attack unacceptably costly. It is reasonable to argue that the absence of Soviet fears is not due to incomplete evidence. The most critical facilities were located close to Soviet borders, making the potential attack much less challenging for the Soviets. The most sophisticated Chinese air defense systems were of Soviet design, making it easier for the Soviets to defeat them with intimate knowledge of their technology, especially considering the widespread Soviet practice of selling downgraded versions of Soviet weapons to foreign countries. Also, the mixed performance of Chinese antiaircraft units in Vietnam does little to make one believe in their ability to inflict prohibitive loses on the Soviets, particularly considering the generally smaller Soviet sensitivity to its own causalities.[132]

To the vital importance for the development of deterrence theory, the summarization of this case's empirical results in Table 3.1 highlights some remarkable similarities to the previous case. Despite being more developed than a few years earlier, China's nuclear arsenal of some 50 nuclear bombs did not constitute a threat of sufficient robustness to deter Soviets as it failed to meet the second strike threshold. But the Soviet military juggernaut was restrained in its first strike contemplation by the possibility of being dragged into a long conventional war in the region, where its conventional army was weaker and harder to supply with insufficient and vulnerable supply lines. Certainly the threat of China's eventual conventional retaliation was not the only limiting factor for Soviets and, quite importantly, the influential "peace party" around Kosygin balanced the hawks in Moscow. But it appears it was the most important one and, by far, more important than the threat of nuclear retaliation.

Table 3.1 USSR–China, 1966

Concept	Description	Values
The deterrer's nuclear arsenal	Key physical attributes of respective nuclear complexes; weapon types; numbers; delivery vehicles; command, control, and communication systems	Approx. 50 (both fission and fusion weapons tested and deployed; ballistic missiles tested but probably not yet deployed; up to 150 Il-28 bombers available; no early warning system; primitive command and control)
Nuclear asymmetry	Situation when a substantially smaller (quantitatively) and less sophisticated nuclear force (qualitatively) faces a qualitatively and quantitatively larger nuclear force	Strong asymmetry (quantitative superiority by a factor of at least one-hundred and comparable qualitative edge)
Second-strike criterion	Nuclear posture that has the ability to survive enemy attack, make and communicate decision to retaliate, overcome enemy's active defense, and destroy a valuable target despite its passive defense	None (loopholes in survivability; imperfect C3 systems; inadequate delivery vehicles available)
General conventional preponderance	Situation when one side's armed forces are in general substantially stronger in terms of numbers, technology, training, and employment strategies	Challenger (stronger navy, air force, and land forces; edge in terms of quality and quantity)
Theater conventional preponderance	Situation when one side's armed forces are substantially stronger in terms of numbers, technology, training, and employment strategies on a theater where targets valuable to the challenger can be found	Deterrer (immediate superiority in numbers; the challenger's vulnerable lines of communication)

continued

Table 3.1 Continued

Concept	Description	Values
The challenger's technological advantage	Significant advantage of the challenger's major weapons systems that would be employed in case of conflict in terms of state-of-the-art sophistication over the deterrer's weapons systems	Strong (significant edge in sophistication of most weapon platforms; some technologies completely unavailable to the deterrer)
Availability of information	The challenger's knowledge about targets' location and defensive systems, which is established from sources that the challenger deems credible	Strong (confidence in own intelligence services, likely insider contacts established during previous cooperation; imagery intelligence available)
Centrality of conflict	Absolute importance of the conflict dyad between the challenger and the deterrer from the challenger's subjective perspective and its relative importance to other existing conflict dyads where the challenger is a party	Yes (attention to the conflict paid by the top decision makers, yet competing for attention with other conflict dyads)
Perceived resolve of the deterrer	The challenger's perception of the deterrer's commitment to fight should the deterrence fail	Strong (demonstrated during wars with Japan, nationalists, and in Korea; vocal disrespect of U.S. nuclear threats)
Institutionalization of mutual relations	Degree of shared expectations about the requirements of stable deterrence and the existence of proven formal or informal communication channels	Strong (formal and informal communication channels available; history of close contacts; yet large number of misunderstandings)
History of hostility	Track record of conflict in the dyad that shapes the understanding of other side's intentions	Limited (some history of border incidents, but overall history of recent cooperation)

Last resort	Situation when the challenger sees only the options to strike, or to live up to the development he tries to prevent	Limited (alternatives to war seen by doves; hawks considered situation advantageous for strike but not the last one)
Perception of the deterrer's rationality	The challenger's perception of the deterrer's rationality, particularly whether the challenger believes it possible to live with the nuclear-armed deterrer in the long term	No (Mao's paper tiger rhetoric on nuclear threats and allegedly irresponsible behavior)
Third-party deterrence	Threat of military involvement into the original conflict by a third party, most likely the allies of the original deterrer, that decisively influenced the challenger's decision	Limited (voiced U.S. disapproval but with little credibility of meaningful action)
Alliance politics	Sensitivity of the challenger to possible impact of his action on the relations with his allies	None (allies noncommittal, but not opposing)
International action legitimacy	The challenger's sensitivity to the international normative expectation of non-intervention	Limited (concerns about action legitimacy not raised)
Nuclear taboo	Normative prohibition on the use of nuclear weapons	Emerging (nuclear first strike opposed by majority in the Politburo, but less so on normative grounds)
Domestic action legitimacy	Level of support for military solution among the challenger's population	Likely not applicable (no concerns about domestic legitimacy raised)
Opposition among influential decision makers	Negative view on military solution by influential part of the challenger's government	Strong (Kosygin; part of KGB, military intelligence; foreign service)
Regime type	The challenger's position on the democracy-nondemocracy axis and on the militarized-nonmilitarized axis	Non-democracy, strong party control of armed forces

continued

Table 3.1 Continued

Concept	Description	Values
Nuclear retaliation	Threat that the deterrer will use its nuclear weapons against targets valuable to the challenger, except of targets that are directly related to pursuit of the challenger's objectives	Limited (great confidence in a first strike)
Nuclear denial	Threat that the deterrer will use its nuclear weapons against targets that are directly related to pursuit of the challenger's objectives in order to prevent him from attaining the objectives, or in order to make it unacceptably costly	Limited (no concerns about China's nuclear denial raised; no preparatory measures taken to mitigate such threat)
Conventional denial	Threat that the deterrer will use its conventional weapons against targets that are directly related to pursuit of the challenger's objectives in order to prevent him from attaining the objectives, or in order to make it unacceptably costly	Limited (no concerns raised; China's complexes close to Soviet border and defended with older Soviet weapons which the USSR likely knew how to defeat; smaller Soviet sensitivity to own causalities)
Conventional retaliation	Threat that the deterrer will use its conventional weapons against targets valuable to the challenger, except of targets that are directly related to pursuit of the challenger's objectives	Strong (concerns about PRC action against Soviet Far East and need to tie down forces in protracted conventional war)

Notes

1 Odd Arne Westad, "Introduction" in Odd Arne Westad (ed.), *Brothers in Arms: The Rise and Fall of the Sino-Soviet Alliance, 1945–1963* (Stanford: Stanford University Press, 1998) 5–12.

2 Viktor M. Gobarev, "Soviet Policy Toward China: Developing Nuclear Weapons 1949–1969," *Journal of Slavic Military Studies*, 12/4 (1999) 4.

3 See Xiaobing Li, *History of Modern Chinese Army* (Lexington: University Press of Kentucky, 2007) 113–146; Sergei Goncharenko, "Sino-Soviet Military Cooperation," in Odd Arne Westad (ed.), *Brothers in Arms: The Rise and Fall of the Sino- Soviet Alliance, 1945–1963* (Stanford: Stanford University Press, 1998) 141–164; You Ji, "The Soviet Model and the Breakdown of Military Alliance," in Thomas P. Bernstein and Hua-yu Li (eds), *China Learns from the Soviet Union: 1949-Present* (Plymouth: Lexington Books, 2010) 131–149.

4 Lorenz M. Luthi, *Sino-Soviet Split: Cold War in the Communist World* (Princeton: Princeton University Press: 2010) 37.

5 Lorenz M. Luthi, "Sino-Soviet Relations during the Mao Year, 1949–1969" in Thomas P. Bernstein and Hua-yu Li (eds), *China Learns from the Soviet Union: 1949–Present* (Plymouth: Lexington Books, 2010) 28.

6 Cheng Jian and Yang Kuisong, "Chinese Politics and the Collapse of the Sino-Soviet Alliance," in Odd Arne Westad (ed.), *Brothers in Arms: The Rise and Fall of the Sino-Soviet Alliance, 1945–1963* (Stanford: Stanford University Press, 1998) 250–255.

7 Westad, "Introduction," 14–15.

8 Zhihua Shen and Yafeng Xia, "Between Aid and Restriction The Soviet Union's Changing Policies on China's Nuclear Weapons Program 1954–1960," *Asian Perspective*, 36/1 (April 2009) 101; Westad, "Introduction," 15.

9 Chen Jian, *Mao's China and the Cold War* (Chapel Hill: University of North Carolina Press, 2001) 62.

10 Shen and Xia, "Between Aid and Restriction," 103.

11 Shen and Xia, "Between Aid and Restriction," 104.

12 Westad, "Introduction," 16.

13 Li, *History of Modern Chineese Army*, 154; Luthi, *Sino-Soviet Split*, 39.

14 Jian, *Mao's China and the Cold War*, 64–65.

15 Westad, "Introduction," 18.

16 Luthi, *Sino-Soviet Split*, 46.

17 Luthi, "Sino-Soviet Relations during the Mao Year, 1949–1969," 34–39.

18 Jian, *Mao's China and the Cold War*, 74.

19 Luthi, *Sino-Soviet Split*, 93.

20 Luthi, *Sino-Soviet Split*, 99; also see Michael M. Sheng, "Mao and China's Relations with the Superpowers in the 1950s: A New Look at the Taiwan Strait Crises and the Sino-Soviet Split," *Modern China*, 34/4 (October 2008) 477–507.

21 Richard Betts, *Nuclear Blackmail and Nuclear Balance* (Washington D.C.: Brookings Institution Press, 1987) 66–72.

22 Odd Arne Westad, "The Sino-Soviet Alliance and the United States," in Odd Arne Westad (ed.), *Brothers in Arms: The Rise and Fall of the Sino-Soviet Alliance, 1945–1963* (Stanford: Stanford University Press, 1998) 176.

23 Shu Guang Zhang, "Between 'Paper' and 'Real' Tigers: Mao's View of Nuclear Weapons," in John Lewis Gaddis, Phillip H. Gordon, Ernest R. May and Joathan Rosenberg (eds), *Cold War Statements Confront the Bomb: Nuclear Diplomacy Since 1945* (New York: Oxford University Press, 1999) 208.

24 Luthi, *Sino-Soviet Split*, 137; Shen and Xia, "Between Aid and Restriction," 108–112.
25 Luthi, *Sino-Soviet Split*, 163.
26 Westad, "Introduction," 25.
27 Luthi, *Sino-Soviet Split*, 174–180.
28 Westad, "Introduction," 27.
29 Luthi, *Sino-Soviet Split*, 200.
30 See previous chapter.
31 Mingjiang Li, "Ideological Dilemma: Mao's China and the Sino-Soviet Split, 1962–1963', *Cold War History*, 11/3 (August 2011) 387–419.
32 Krista E. Wiegand, *Enduring Territorial Disputes: Strategies of Bargaining, Coercive Diplomacy, and Settlement* (Athens GA: University of Georgia Press, 2011) 226–230.
33 Luthi, *Sino-Soviet Split*, 181.
34 Wiegand, *Enduring Territorial Disputes*, 231.
35 Luthi, *Sino-Soviet Split*, 213–217.
36 Wigand, *Enduring Territorial Disputes*, 231.
37 Nicholas Khoo, *Collateral Damage: Sino-Soviet Rivalry and the Termination of the Sino-Vietnamese Alliance* (New York: Columbia University Press, 2011) 16–17.
38 Khoo, *Collateral Damage*, 20; Luthi, *Sino-Soviet Split*, 288–291.
39 Khoo, *Collateral Damage*, 23–24.
40 Khoo, *Collateral Damage*, 58; Yang Kuisong, "The Sino-Soviet Border Clash of 1969: From Zhenbao Island to Sino-American Rapprochement," *Cold War History*, 1/1 (August 2000) 24.
41 William A. Joseph. "Foreword" in Gao Yuan, *Born Red: A Chronicle of the Cultural Revolution* (Palo Alto: Stanford University Press, 1987) xiv.
42 Matthew J. Ouimet, *Rise and Fall of the Brezhnev Doctrine in Soviet Foreign Policy* (Chapel Hill: University of North Carolina Press, 2003) 67.
43 Khoo, *Collateral Damage*, 47–49.
44 Luthi, *Sino-Soviet Split*, 341. At full strength, Soviet divisions had between 7,000 to 10,500 men, while Chinese divisions were, on average, 2,000 to 4,000 men stronger; see Lyle L. Goldstein, "Return to Zhenbao Island: Who Started Shooting and Why It Matters," *The China Quarterly*, 168 (December, 2001) fn. 38, fn. 39; overall, the number of the Chinese troops is estimated around 400,000; William H. Mott and Jae Chang Kim, *Philosophy of Chinese Military Culture: Shih vs. Li* (Gordonsville: Palgrave Macmillan, 2006) 169.
45 Goldstein, *Preventive Attack and Weapons of Mass Destruction*, 77–78; Kuisong, "The Sino-Soviet Border Clash of 1969," 27–31; Wiegand, *Enduring Territorial Disputes*, 236; Khoo, *Collateral Damage*, 58.
46 Kuisong, "The Sino-Soviet Border Clash of 1969," 19; William H. Mott and Jae Chang Kim, *Philosophy of Chinese Military Culture: Shih vs. Li*, 173.
47 For various divergent and incomplete pictures of the two incidents, see Lyle J. Goldstein, *Preventive Attack and Weapons of Mass Destruction: A Comparative Historical Analysis* (Stanford: Stanford University Press, 2006) 77–78.
48 Goldstein, "Return to Zhenbao," 997; Kuisong, "The Sino-Soviet Border Clash of 1969," 27–31; an official Russian version witch appears to be less burdened by propaganda can be retrieved from a Soviet memorandum to GDR leadership, see Christian F. Ostermann, "New Evidence on The Sino-Soviet Border Dispute, 1969–1971," *Cold War International History Project Bulletin*, 6–7

(Winter 1995–1996) 186–191; an interesting Russian description can also be obtained at www.damanski-zhenbao.ru/chronicals_en.htm

49 Kuisong, "The Sino-Soviet Border Clash of 1969," 27–31.

50 Goldstein, *Preventive Attack and Weapons of Mass Destruction*, 81.

51 Arkady L. Schevchenko, *Breaking with Moscow* (New York: Alfred A. Knopf, 1985) 165.

52 Vladislav M. Zubok, *A Failed Empire: The Soviet Union in the Cold War from Stalin to Gorbachev* (Chapel Hill: The University of North Karolina Press, 2007) 201.

53 Luthi, *Sino-Soviet Split*, 342.

54 Osterman, "New Evidence on The Sino-Soviet Border Dispute, 1969–71," 188

55 Kuisong, "The Sino-Soviet Border Clash of 1969," 32.

56 Mott and Kim, *Philosophy of Chinese Military Culture*, 177.

57 Goldstein, *"*Return to Zhenbao," 992.

58 Gobarev, "Soviet Policy toward China," 46.

59 Gobarev, "Soviet Policy toward China," 40.

60 Goldstein, *Preventive Attack and Weapons of Mass Destruction*, 84.

61 Lewis, *The Minimum Means of Reprisal*, 15.

62 Goldstein, *Preventive Attack and Weapons of Mass Destruction*, 78–79.

63 Luthi, *Sino-Soviet Split*, 342; William Burr, "Sino-American Relations, 1969: The Sino-Soviet Border War and Steps Toward Rapprochement," *Cold War History*, 1/3 (2001) 85.

64 Kuisong, "The Sino-Soviet Border Clash of 1969," 34.

65 Chen Jian and David L. Wilson, "'All Under the Heaven is a Great Chaos': Beijing, the Sino-Soviet Clashes and the Turn Toward Sino-American Rapprochement, 1968–1969," *Cold War International History Project Bulletin*, 11 (Winter 1998) 168–169.

66 Mott and Kim, *Philosophy of Chinese Military Culture*, 178.

67 Ostermann, "New Evidence on The Sino-Soviet Border Dispute, 1969–1971," 192.

68 Kuisong, "The Sino-Soviet Border Clash of 1969," 38.

69 Goldstein, *Preventive Attack and Weapons of Mass Destruction*, 80.

70 Goldstein, "Return to Zhenbao Island," 985.

71 Kuisong, "The Sino-Soviet Border Clash of 1969," 40–41.

72 See previous chapter.

73 Hans M. Kristensen and Robert S. Norris, "Global Nuclear Weapons Inventories, 1945–2013," *Bulletin of the Atomic Scientists*, 69/5 (2013) 78.

74 John W. Lewis and Xue Litai, *China Builds the Bomb* (Stanford: Stanford University Press, 1991) 214.

75 Goldstein, *Preventive Attack and Weapons of Mass Destruction*, 84–85.

76 Mott and Kim, *Philosophy of Chinese Military Culture*, 167; Goldstein, *Preventive Attack and Weapons of Mass Destruction*, 85.

77 John L. Gaddis, *Strategies of Containment: A Critical Appraisal of American National Security Policy During the Cold War* (Cary: Oxford University Press, 2005) 284.

78 Robert S. Norris and Hans M. Kristensen, "Nuclear U.S. and Soviet/Russian Intercontinental Ballistic Missiles, 1959–2008," *Bulletin of the Atomic Scientists*, 65 (January/February 2009) 70; Pavel Podvig, "The Window of Vulnerability That Wasn't: Soviet Military Buildup in the 1970s: A Research Note," *International Security*, 33/1 (Summer 2008) 125.

79 Pavel Podvig, *Russian Strategic Nuclear Forces* (Cambridge: MIT Press, 2004) 339–347.

80 Pavel Podvig, "History and the Current Status of the Russian Early-Warning System," *Science and Global Security*, 10/1 (2002) 27, 41.

81 Chen Jian and David L. Wilson, "'All Under the Heaven is a Great Chaos': Beijing, the Sino-Soviet Clashes and the Turn Toward Sino-American Rapprochement, 1968–1969," *Cold War International History Project Bulletin*, 11 (Winter 1998) 173.

82 Goldstein, *Preventive Attack and Weapons of Mass Destruction*, 80.

83 Goldstein, *Preventive Attack and Weapons of Mass Destruction*, 86–87.

84 Goldstein, "Return to Zhenbao," 993.

85 Xiaobing Li, *History of the Modern Chinese Army* (Lexington: University Press of Kentucky, 2007) 234.

86 Li, *History of the Modern Chinese Army*, 256.

87 Kuisong, "The Sino-Soviet Border Clash of 1969," 26–27.

88 Gaddis, *Strategies of Containment*, 284.

89 Luthi, *Sino-Soviet Split,* 344.

90 Lyle J. Goldstein, "Do Nascent WMD Arsenals Deter?: The Sino Soviet Crisis of 1969," *Political Science Quarterly*, 118/1 (Spring 2003) 59.

91 Luthi, *Sino-Soviet Split*, 340.

92 Goldstein, *Preventive Attack and Weapons of Mass Destruction*, 91.

93 Luthi, *Sino-Soviet Split*, 340.

94 Goldsteim, "Return to Zhenbao," 992.

95 Luthi, *Sino-Soviet Split*, 197; Paul Jackson (ed.), *Jane's All the World's Aircrafts* (Alexandria: Jane's Information Group Limited, 2004) 77.

96 Goldstein, *Preventive Attack and Weapons of Mass Destruction*, 68.

97 Goldstein, *Preventive Attack and Weapons of Mass Destruction*, 87.

98 Jian, *Mao's China and the Cold War*, 242.

99 Zubok, *A Failed Empire*, 210.

100 Goldstein, *Preventive Attack and Weapons of Mass Destruction*, 81.

101 Goldstein, *Preventive Attack and Weapons of Mass Destruction*, 80.

102 Luthi, *Sino-Soviet Split*, 342.

103 Wiegand, *Enduring Territorial Disputes*, 228.

104 Wiegand, *Enduring Territorial Disputes*, 226.

105 Goldstein, *Preventive Attack and Weapons of Mass Destruction*, 82.

106 Luthi, *Sino-Soviet Split*, 77.

107 Luthi, *Sino-Soviet Split*, 77.

108 Luthi, *Sino-Soviet Split*, 101.

109 Zhang, "Between 'Paper' and 'Real' Tigers," 194–215.

110 Burr, "Sino-American Relations, 1969," 75–80.

111 Betts, *Nuclear Blackmail and Nuclear Balance*, 81.

112 Goldstein, *Preventive Attack and Weapons of Mass Destruction*, 89.

113 Mott and Kim, *Philosophy of Chinese Military Culture*, 175.

114 Luthi, *Sino-Soviet Split*, 343.

115 Ouimet, *Rise and Fall of the Brezhnev Doctrine in Soviet Foreign Policy.*

116 Nina Tannewald, "The Nuclear Taboo: The United States and the Normative Basis of Nuclear Non-Use," *International Organization*, 53/3 (Summer 1999) 464; Nina Tannenwald, *The Nuclear Taboo: The United States and the Non-Use of Nuclear Weapons Since 1945* (Cambridge: Cambridge University Press, 2007) 368.

117 Gobarev, "Soviet Policy toward China," 46.

118 Based on Gobarev, "Soviet Policy toward China," 38–41.

119 Khoo, *Collateral Damage*, 17.

120 Zubok, *A Failed Empire*, 210.
121 Zubok, *A Failed Empire*, 201.
122 Wiegand, *Enduring Territorial Disputes*, 248.
123 Brian D. Taylor, *Politics and the Russian Army: Civil–Military Relations, 1689–2000* (Cambridge: Cambridge University Press, 2003) 200.
124 Taylor, *Politics and the Russian Army*, 190–205.
125 Interestingly, the Soviets were well aware of the plausibility of nuclear denial strategies and reportedly considered their own nuclear denial for dealing with China when the KGB proposed using nuclear land mines along the Sino-Soviet border. See Gobarev, "Soviet Policy toward China," 40.
126 Kuisong, "The Sino-Soviet Border Clash of 1969," 36.
127 Jeffrey Lewis, *The Minimum Means of Reprisal: China's Search for Security in Nuclear Age* (Cambridge: MIT Press, 2007) 15.
128 Goldstein, *Preventive Attack and Weapons of Mass Destruction*, 86–87.
129 Wiegand, *Enduring Territorial Disputes*, 242.
130 Gobarev, "Soviet Policy toward China," 40; Zubok, *A Failed Empire*, 210.
131 Goldstein, *Preventive Attack and Weapons of Mass Destruction*, 80.
132 On the performance of PRC's air defense units in Vietnam, see Li, *History of the Modern Chinese Army*, 215–226; and Xiaoming Zhang, "The Vietnam War, 1964–1969: A Chinese Perspective," *The Journal of Military History*, 60/4 (October, 1996) 754–759.

4   Israel and Iraq, 1977–1981

The 1981 destruction of a French-supplied and almost completed Iraqi nuclear reactor, usually known as Osiraq, offers in a single package a case that is simultaneously invaluable and incredibly prone to criticism. On the one hand, Osiraq is a rare case of deterrence failure in what was almost a nuclear dyad, allowing scholars to observe deterrence processes. Also, despite the characteristics of Israel's democracy – whose military censor jealously guards disclosing certain national security issues, most prominently the existence of Israel's nuclear arsenal – otherwise liberal policies, free press, and frequent leaks from the government make empirical evidence reasonably available. However, it must be admitted that the case has its downsides. Considering Iraq in 1981 a small nuclear state requires very relaxed criteria. Even accepting that, due to uncertainty at the latter stage of nuclear weapons development, progress is often impossible to determine for outsiders, making the difference between zero and a few essentially irrelevant, helps only a little. Yet the case still deserves examination. First, Iraq already had the means to retaliate, including SCUD missiles and 12 kg of French HEU, which some claim was enough for one bomb. Second, and even more importantly, there are striking similarities with the other cases, to be highlighted in the cross-case comparison.

The plot

Iraq had already embarked on the nuclear journey in the 1950s, but, as with most Third World countries that started to work with nuclear technologies at the time, Iraq had little clue as to what the goal was.[1] In 1956, the Iraqi government initiated a nuclear program under the Atoms for Peace umbrella.[2] Three years later, the Iraqi Atomic Energy Commission was established and subsequently a two thermal megawatt (MWt) research reactor was obtained from the USSR. The reactor went critical in 1967 and Iraq signed the NPT in 1968 and ratified it a year later.[3] At the time, there were few reasons to worry. The Soviet-supplied reactor was too small to produce a sensible amount of bomb-suitable material. Furthermore, the Soviets – whose tough nonproliferation policy is often forgotten, probably thanks to Western Cold War propaganda – were happy to charge Iraq badly needed hard currency for nuclear cooperation, but they had no intention of letting the Iraqis arm with nuclear weapons. Soviet personnel carefully watched over the reactor, giving the Iraqis little chance to divert from its declared use. It appears that, in the course of the reactor's lifetime, the Iraqis decided not to pay for overpriced Soviet maintenance. Yet due to the lack of indigenous nuclear reactor maintenance know-how, the Iraqis damaged the reactor's core while it was being cleaned by a local industrial cleaning service and subsequently the reactor was phased out.[4]

The pace of the Iraqi nuclear program significantly accelerated when Saddam Hussein appointed himself head of the Iraqi Atomic Energy Commission.[5] Yet Saddam still probably lacked a clear vision on what to do with a nuclear program, much less how to do it, and assigned little urgency to the issue. Iraqi scientists believed that the breakout option was preferred by their leader, though it does not seem they had an explicit authorization to head in that direction. The program was moving slowly toward its presumed goal, struggling with inadequate funding, staff, mandate, and infrastructure.[6]

Despite the lack of clear goals, Saddam turned to the French and Italians for help with his nuclear program. The recent oil crisis made both willing to cooperate. To the liking of then French President Giscard D'Estaing and his Prime Minister Chirac, the latter's visit to Bagdad in 1974 and two following visits of Saddam to Paris yielded numerous contracts on the delivery of French industrial equipment – unsurprisingly including French weapons that were high on Saddam's wish list – and on deliveries of Iraqi oil for a reasonable price.[7] Early in the negotiations, the Iraqis requested that France supply a 500 MW gas-graphite reactor well suited for plutonium production. Yet the French turned down the request, arguing that the technology was discontinued in production and replaced by more effective systems in the field of electricity generation.[8] As a replacement, the Iraqis opted for an Osiris-type 70 MW (some sources claim 40 MW) material test research reactor together with a smaller 800 kWt Isis reactor. The selection was suspicious as material test reactors are normally designated to test materials to be used in nuclear power stations. Only nations that produce their own nuclear reactors have a good use for a material test reactor. Late 1970s Iraq was certainly not on this rather short list of advanced industrial nations capable of such an endeavor.[9] Soon, numerous objections appeared around the world, including most prominently Israel and the United States, as well as in France itself where André Giraud, head of the French Nuclear Energy Committee, protested together with Ministry of Foreign Affairs officials and reportedly with members of French intelligence services.[10] However, the deal had been forged and was signed in November 1976.

Whether Osiraq was suitable for nuclear weapons production remains an open question. Certainly the reactor's design was suboptimal for the job.[11] To produce plutonium in substantial quantities it would require modifications to its normal operations, namely uranium blankets placed either into the reactor or, less effectively, into the neutron hall underneath. According to the current accounts of Iraq's scientists who worked at the program and to IAEA estimates, should this happen the reactor could produce 2 kg of plutonium per year. Yet the modifications to produce plutonium could be detected by foreign inspections on the spot. IAEA teams arrived in Osiraq for one to two visits a month and the permanent presence of French technicians was written into the Franco-Iraqi treaty, making it difficult to divert material for atomic weapons without detection.[12] Nevertheless this did not prevent Saddam from giving not-too-subtle hints about his ability to have nuclear weapons by 1985.[13]

Israeli intelligence carefully watched the progress of the Iraqi nuclear program, though the threat was not considered imminent at least till 1975.[14] Israeli activity increased after the forging of the Franco-Iraqi deal. The first steps were diplomatic. Foreign Minister Peres, a man who was instrumental in establishing Israel's nuclear program and most importantly its deal with France, turned directly to Chirac, whom he knew well, but with little success. At the same time, Prime Minister Rabin turned to Jewish-American organizations to pressure the U.S. administration to kill the deal France had made with Iraq.[15] Even though Israel's diplomacy pressed the French directly and indirectly, to Tel Aviv it seemed that not much was achieved. In fact, it appeared that matters were becoming even worse. In January 1976, Iraq signed a contract with the SNIA Techint Company of Italy to deliver a facility for plutonium separation known as hot cells. To keep the delivery in compliance with IAEA regulations, only a small "hot version" for study purposes was delivered together with a large one which the Italians labeled as a "training facility" and which lacked biological shielding and tanks which could withstand high irradiation. However, both the Italians and the Iraqis knew that alterations could be made to convert the facility for full-fledged plutonium separation.[16] Also, since 1977 when Iraq started shopping for natural uranium at international markets, it obtained – according to Israeli sources – some 250 tons from Brazil, Portugal, and Niger.[17]

The Israeli pleas and the American pressure were not completely unsuccessful with the French. In late 1970, France developed a new type of fuel for nuclear reactors called "caramel." Caramel fuel was enriched only to 7–9 percent and – contrary to HEU that was supposed to fuel Osiraq according to the original agreement – it was less suitable for plutonium production and unsuitable for direct use in weapons. The French suggested that caramel fuel should be supplied for Osiraq, but the Iraqis insisted on the earlier contract, citing that caramel fuel was not yet in industrial production.[18] The French then decided to deliver the promised HEU to fuel the reactor, but took precautions to mitigate Iraqi chances of misusing it. Reportedly, the French pre-irradiated the fuel to be supplied, making manipulation with it hazardous.[19] Paris also made the Iraqis agree that instead of 70 kg of HEU as promised in original agreement, Iraq would get only 12 kg for Osiraq and 12 kg for Isis.[20] The first 12 kg were delivered in June 1980, but the other batch never arrived.[21]

Iraq's decision to refuse caramel fuel only added to existing doubts about its intentions, most profoundly in Israel. Tel Aviv had at least five indications to believe that Saddam's intention was not peaceful use of nuclear energy: the initial request for a gas-graphite reactor; the decision to buy a material test reactor; the demand for natural uranium; the acquisition of hot cells; and now the refusal to accept the caramel fuel instead of the weapon-grade HEU.[22] Few in Israel doubted Saddam's ultimate goal and few believed that he could be stopped by international safeguards. Not only did the Israelis question the ability of inspections to detect suspicious behavior, but Tel Aviv had little trust in the ability of the international community to enforce Saddam's compliance in the case of a discovered breach of NPT obligations. Publicly, Israel referred to the inability of the international community to deal with Pakistan's and Libya's known misbehavior.[23] Privately, Israeli leaders knew well from their own experience that others could be fooled and the bomb could be built under the cover of secrecy.[24]

According to Claire, shortly after the 1977 elections, the newly elected Israeli Prime Minister Menecham Begin summoned his security policy inner circle to address what was perceived to be a growing threat to Israel.[25] The reliability of this information is however uncertain as Claire appears to be giving confusing information, stating at one instance that Begin was present as the Prime Minister, while attributing the same role to Rabin at another instance. Claire also claims that Yehushua Saguy attended the meeting as the head of AMAN, Israel's military intelligence; however, Saguy became the head of AMAN only in 1979. The more reliable Perlmutter, Handel, and Bar-Joseph state that the meeting took place on August 23, 1978, and was attended by Prime Minister Begin, Deputy Prime Minister Yigael Yadin, Foreign Minister Moshe Dayan, Defense Minister Ezer Weitzman and his deputy, Mordechai Zipori, Agriculture Minister Ariel Sharon, Finance Minister Simha Ehrlich, and two additional ministers together with the Chief of the IDF (Israel Defense Force) Staff General Raphael Eitan, AMAN's head Shlomo Gazit, his expert on the Iraqi nuclear program Colonel David Bnaya, Mossad Deputy Director Nahum Admoni, Israeli Atomic Energy Commission Director Uzi Eilam and a "number of personal assistants."[26] Notwithstanding the actual date and participation, some important accounts of the meeting appear undisputed. First, there was an agreement that Iraqi nuclear ambitions had to be stopped. Second, the participants disagreed on how to do it. Begin, Sharon, and Eitan, among others, supported the use of military means, while Yadin and Weitzman opposed it, arguing that the threat was not urgent and, at the moment, the political costs would outweigh the benefits of the strike.

The IDF had already prepared some initial contingency plans before the meeting. Afterwards, the military was instructed to elaborate detailed plans for an operation against Osiraq. In the meantime, diplomatic efforts continued. In February 1979, Foreign Minister Dayan visited Paris, but failed to convince the French to stop the cooperation with the Iraqis.[27] Soon, other options to hamper Baghdad's nuclear effort were added to the continuing calls of Israel's diplomacy. On April 6, 1979, an explosion in the CNIM warehouse at the port of Le Seyne-sur-Mer in the Toulon agglomeration hit the cores of two reactors prepared for delivery to Iraq.[28] Just a month later, the dead body of Dr. Yahya El Meshad was found in a hotel in Paris. Egyptian by birth and a top-class nuclear scientist by training, El Meshad was working for Saddam's nuclear program and had arrived in Paris to oversee the delivery of the first 12 kg of HEU for Iraqi reactors.[29] The next year, bombs exploded in the apartment of SNIA Techint General Manager and in the company's offices. This time no one was hurt, but the explosions were followed by a call from a completely unknown terrorist organization named "The Committee to Safeguard the Islamic Revolution" which warned against cooperation with Iraq.[30] Investigations into these actions were unsuccessful, but the patterns are suggestive. Previously unknown groups claimed responsibility for the attack and neither of the groups (the French Ecological Group claimed responsibility for the Le Seyne-sur-Mer bombing) attacked again, casting doubts on their existence. The crimes were committed by professionals; apparently Saddam's nuclear program was to be hampered as much as possible but Western casualties were carefully avoided.[31] Israel's famous secret service, Mossad, topped the list of suspects. But the list was quite long – the United States, the USSR, French intelligence, Iran, Syria all had both good reasons and sufficient capacities to commit one or all of the acts – and evidence was scarce.

According to information from Khidir Hamza, Saddam directly instructed his scientists in December 1979 to build the bomb.[32] Again, the reliability of this information is unclear. The decision would have made sense, as Saddam's fears were growing after the Iranian revolution. But in December 1979, Saddam also ordered the arrest of one of his nuclear program's top scientists, Hussain al-Shahristani, for alleged support of an illegal Shiite movement. Just a month later, Jafar Dhia Jafar, his boss and another top nuclear scientist, was also arrested after protesting over Shahristani's fate. The real motivation for Shahristani's arrest may have been his reaction to Saddam's request to work on a nuclear bomb.[33] In any case, while the Iraqi leader may have had immense oil wealth, he certainly did not have a great number of top nuclear scientists. The absence of Jafar and Shahristani did little to help Saddam's nuclear ambitions. In June 1980, both imprisoned nuclear scientists were informed that Saddam wanted a nuclear weapon and apparently offered incentives to cooperate. Shahristani refused and remained in prison, while Jafar conditionally agreed and was transferred to house arrest; he only returned to Iraq's nuclear program after the Israeli raid.[34]

Whether Saddam explicitly instructed his scientists to build the bomb in December 1979 or July 1980 is uncertain, though the earlier date is more likely. Iraqi records captured after the American overthrowing of Saddam's regime reveal that in March 1979 Saddam told high-level Iraqi officials that Iraq would get nuclear weapons from its "Soviet friends."[35] But he could hardly hope for that; the USSR was happy to supply Iraq with Soviet weapons, usually with downgraded export models, nonetheless by 1979 Saddam must have known well that Soviet nuclear weapons were not for sale. Moscow did not want to sell Saddam anything bigger than a small research reactor. Turning to his own scientists was a reasonable option.

The reasons for an explicit order to make nuclear weapons are even less clear. The captured Iraqi records show that Saddam considered balancing Israel's nuclear arsenal to be the highest priority for his nuclear weapons program. The bomb was to allow the Arab coalition to fight a limited war of attrition against Israel without being subjected to Israeli nuclear retaliation and to reconquer the lands lost in 1967.[36] If so, the timing would be probably well explained by Egypt's peace treaty with Israel. Saddam's aforementioned reference to his ability to get bomb from the Soviets dates March 27, 1979, one day after the peace treaty was signed in Washington. However, equally important for the Iraqi leader may have been the threat of the 1979 Iranian revolution.[37]

The disagreements between the two countries turned into a bloody war when Saddam's army invaded Iran in September 1980. In a series of retaliatory raids that the Iranian air force conducted against various targets in Iraq in late September, two Iranian F-4 Phantoms attacked the Osiraq reactor, but caused only minor damage.[38] From Israel's perspective, the unsuccessful Iranian strike did little good. Iraqi anti-aircraft defense around Osiraq, which failed completely to respond to the Iranian attack, was substantially increased. Also worrisome for Tel Aviv was the expulsion of foreign technicians from the Al Tuwaitha nuclear complex and the refusal to let IAEA inspectors in, a decision that the Iraqis justified on the grounds of the necessity to ensure their safety.[39] By spring 1981, when foreigners returned, Israeli planning for the destruction of Osiraq was already completed.

By February 1980, Israeli Air Force (IAF) commander General David Ivry presented Minister Weitzman and General Eitan with options. The airstrike was selected as a preferred one.[40] It was apparent that the mission would be neither easy nor ordinary. Israel's air force was among the world's strongest, both in terms of training and technology and in terms of numbers. Furthermore, Israel had an abundance of pilots with real combat experience; in the group of eight pilots that bombed Osiraq, only future astronaut Ilan Ramon would be on his first combat mission. But IAF pilots were used to flying only short missions over their small country and in its close vicinity, while Bagdad was 500 miles away. And 500 miles was a direct route with unacceptable coverage by radar and a hostile air force. It was decided that in order to avoid detection, the strike package would flight in low attitude from the Etzion airbase in the Sinai across the Gulf of Aqaba, southern Jordan, and then across northern Saudi Arabia. Yet the decision had its tradeoffs. The distance would rise to 600 miles and the low flight would burn more fuel. It was uncertain whether the aircraft would have enough fuel for the way back.

Selecting the right aircraft for the mission was an uneasy task. IAF's standard fighters – A-4 Skyhawks and indigenous Kfirs – lacked sophisticated radar and bombing systems and would need inflight refueling. The aged F-4 Phantoms were regularly upgraded and could use early precision-guided munition (PGM), but the bulky planes lacked maneuverability and required a two-man crew. Furthermore, the reliability of the early PGM was uncertain. The best solution appeared to be Israel's newest American-made fighter, the F-15, but that required special fuel tanks, which the U.S. were reluctant to provide, arguing that Israel did not need a longer range for its defense. Yet fortune smiled on the Israeli planners. Shortly after the Shah of Iran was overthrown, the U.S. canceled the sale of their newest F-16s – introduced only in 1978 – which the Shah had ordered for his air force. Eight were already assembled and General Dynamics was eager to find a new buyer. Israel gladly accepted them, even though the F-16 lacked the sufficient range. The IAF planners found a risky solution. The attacking aircraft would do a "hot refueling" before takeoff and drop external fuel tanks to reduce their weight.[41] Neither of the procedures was in compliance with the plane's manual.[42]

In October 1980, Mossad reported to the Prime Minister that the Osiraq reactor would be hot by June 1981.[43] Begin decided that Osiraq had to be destroyed before that happened. Subsequently, the issue was discussed first by Begin's "security cabinet" on September 14, and by the full cabinet two weeks later. Apparently, the top decision makers were divided and Begin's position faced substantial opposition. Both the heads of his intelligence services – Mossad's Yatzak Hofi and AMAN's new chief Yehoshua Saguy – objected together with Deputy Prime Minister Yadin. Previously one of the most vocal critics, Defense Minister Weitzman, did not object but only because he had already resigned in May in protest against the plan. While no one questioned the wisdom of keeping nuclear weapons out of Saddam's hands, the critics argued that time was still available for other ways to do it.

Those not favoring the strike also pointed out the likely negative consequences on the relations with the United States, on the fresh peace process with Egypt, as well as the possibility of reuniting Arabs against Israel. The threat of enhancing the Soviet Union's position in the region was also raised along with the possibility that the attack would only press the Arabs to step up their efforts to get the bomb.[44] Altogether, six of the 16 ministers refused to support the decision, with Yadin threatening his resignation.[45] But the green light was given anyway as the government voted to leave the decision about the timing to the "security cabinet."[46]

The security cabinet met on May 3, 1981, and a majority of the ministers approved the strike. The date was set for May 10.[47] The planes were ready to take off on May 10, all pilots informed about their target, but the strike was canceled at the last moment.[48] Opposition leader Shimon Peres had learned the date of attack from former Defense Minister Weitzman. Subsequently he addressed a cryptic latter to Begin, urging him to postpone the strike.[49] Peres warned against rushing into the strike and about its consequences, but one thing probably preoccupied him the most. French presidential elections were scheduled for May 10, 1981. Peres's friend François Mitterrand was running as the socialist candidate against Giscard. Previously, Mitterrand had assured Peres that France under his leadership would not supply Iraq with HEU.[50] Begin was reportedly furious that the date had been revealed, but canceled the strike. Most sources claim that the disclosure of the date made Begin cancel and decide for another date, but this may have been only Begin's explanation to others, while in fact he decided to wait for the outcome of the French elections.[51] At the end of the day, waiting any longer, with more people knowing about the operation, increased the chance that the Iraqis would learn.

The Israeli strike did not take place on May 10, but François Mitterrand was elected President of France. Shortly after the elections, new Foreign Minister Claud Cheysson "confirmed that the agreements signed by Giscard's government would be fulfilled."[52] The Israelis understood this to be a turnabout in Mitterrand's policy, meaning that Iraq would get what it was promised, though this is debatable since Cheysson did not explicitly mention the Iraqi nuclear program, and likely referred to the conventional arms sales Mitterrand temporarily suspended after the elections.[53] Yet the destiny of Osiraq was sealed.

On June 7, the Etzion air base was left by eight Israeli F-16s, each carrying two Mk-84 2,000 lb bombs, and eight F-15s designated to provide air cover, jam Iraqi radars over Osiraq, and facilitate radio connection with the F-16s. Two hours later, the reactor was destroyed by 14 bombs. The formidable Iraqi air defense around the nuclear complex was completely taken by surprise. Iraq only learned the identity of the attackers when Begin's government aired the information on Israeli radio.

Unfolding the complexity

As has been noted, the complexity of deterrence failure that culminated in the destruction of Osiraq bears an interesting degree of resemblance to the other cases. But again, treating Iraq as a small nuclear state requires a bit of exaggeration.

From the Israeli point of view, there was little doubt about Saddam's ultimate intention to build the bomb. Iraqi decisions, from the initial attempt to get a large but obsolete gas-graphite reactor to the insistence on fueling Osiraq with HEU, were only too revealing to Tel Aviv. Considering Saddam's offensive rhetoric – including his real and alleged claims such as to be involved in the first attempt at Arab nuclear arming, to use nuclear technology for freeing the Palestine, and his message to Iranians that Osiraq was not aimed against them but against the Zionist entity – it is also little surprise that Israel likely overestimated Iraq's ability to build the bomb.[54] While almost no one believed that Iraq was already nuclear-armed, the intelligence estimated that Iraq had enough plutonium for at least two nuclear bombs.[55] Importantly, apart from Osiraq itself, Saddam had originally been promised 70 kg of HEU to fuel his reactors. This would have sufficed for several bombs, but the French changed the terms of the agreement, thus only 24 kg would be available at any given time and only 12 were actually delivered.[56] If diverted from the reactor, 24 kg of HEU would suffice for at least one bomb, but the experts disagreed as to whether the 12 kilograms that Saddam actually received would be enough for one bomb.[57] Certainly any diversion of HEU from the reactor would be detected, Saddam's malign intention revealed and he would be left with one bomb and no fuel to build another. Only the most optimistic theories, such as postexistential deterrence, expect that nuclear deterrence can work under such circumstances.[58]

Iraq's uncertain ability to get the bomb in the near future is in stark contrast to Israel's unacknowledged, but fairly sophisticated, nuclear arsenal. Israel's policy of opacity makes it impossible to describe its arsenal in detail with confidence, but allows us to sufficiently understand the degree of asymmetry. It is estimated that Israel has deployed 80–200 nuclear weapons. In the early 1980s, Israel had a number of aircraft for the delivery of its nuclear weapons, including the aging F-4 Phantoms, the indigenous Kfir C2s, and the new American F-15s and F-16s.[59] Israel also deployed nuclear-capable Jericho missiles, though their range did not allow for striking Bagdad or other valuable targets in Iraq.[60] While only aircraft were available as delivery vehicles against targets in Iraq, the poor performance of Iraqi air defense vis-à-vis the Israeli strike package at Osiraq, as well as against various Iranian air raids during the Iran–Iraq war, suggests that the Israeli ability to strike Iraq by aircraft would not be hampered.

On the contrary, Iraq's ability to strike Israel if it diverted the HEU to build one bomb would be more theoretical. As the events a decade later revealed, Saddam was able to strike Israel with the SCUD missiles that he had in abundance. He could even strike Israel with chemical weapons, but was reportedly deterred by the fear of Israel's retaliation.[61] Nonetheless, the production of a warhead small enough to fit its SCUDs was hardly within Iraq's capabilities. In such circumstances, Iraq had to turn to its aircraft for delivery. The theoretical power of Iraqi air force was not negligible. It deployed 450 attack aircraft by 1980, of which 140 were high-quality Soviet MIG-23s, SU-7s, and SU-20s and TU-22s.[62] French Mirage F1s which Saddam procured during his negotiations with the French in late 1970s were becoming operational in the Iraqi air force as well. Yet despite the availability of modern technology, the ability of the Iraqi air force to strike targets in Israel in a second strike was doubtful. The experience of two recent Arab-Israel wars was revealing. Only very few Arab aircraft penetrated Israel's airspace in the 1967 and 1973 conflicts as the Israeli air force quickly established its total supremacy over both Israel proper and over the battlefield.[63] Iraq was almost certainly going to lose its only nuclear weapon on the flight.

As the 1980s were approaching, the conventional situation appeared less favorable for Israel. In 1977, the GDP of Iraq reached U.S.$18 billion, approximately 10 times more than Israel's. Iraq also fielded a 190,000 strong army of 12 divisions equipped with 2,200 tanks, including 1,000 modern T-62s, and 1,700 pieces of artillery.[64] Israel was surely at least equal to Iraq with its 169,000 active duty personnel, 3,050 tanks including the high-quality first indigenous Merkavas, and 1,178 pieces of artillery.[65] In the air, Israel enjoyed superiority with 650 planes, mostly Skyhawk A-4, Phantoms, Kfir C-2s, F-15s and F-16s.[66] Furthermore, the IDF certainly enjoyed better tactical abilities and its soldiers were better trained. Yet the rapid growth of Iraqi armed forces got Israel nervous. At the time, Tel Aviv did not consider sole military balance vis-à-vis Iraq relevant. What appeared relevant was the strength of a prospective combined Arab coalition. Iraqi growth was tipping this balance in the late 1970s. Yet two political events ruined the advantage the Arab coalition might have had by 1981. First, in 1979 Anwar Sadat's policy culminated in the peace treaty between Egypt and Israel. Without Egypt's participation, a viable coalition against Israel could be hardly formed.[67] Second, Saddam's invasion of Iran quickly tied Iraq's military power in a long indecisive war on its eastern border.[68]

In contrast to other cases where great powers were involved, both Iraq and Israel were only regional powers, making the borderline between theater and general conventional preponderance blurred. Israel's power projection vis-à-vis Iraq was limited to its air force and so were Israeli ambitions. The only theater relevant for the Israelis was at their borders. On the other hand, Saddam had multiple theaters to think about, and the possibility that a joint

Arab army under his leadership would once reconquer the lost territories from 1967 was one of them.[69] Iraq had participated in previous wars against Israel only to make little difference. However, by 1979, the CIA estimated that Iraq was able to devote a substantial force, some five divisions, to a military conflict with Israel.[70] The reality of this figure was uncertain unless Iraq could secure Syrian help with the logistics, but clearly Saddam contemplated a war at Israel's borders.[71] Either way, by summer 1981, the Iraqi leader desperately needed all the forces he could muster in his ill-fated war with Iran.

Israel certainly enjoyed an advantage in terms of technology and the sophistication of some of its weapons systems. Importantly for the destruction of Osiraq, it already deployed fourth generation fighters, F-15s and F-16s, with modern electronics and avionics, though the bulk of the IAF still comprised of older Phantoms, Skyhawks, Mirage IIIs, and Kfirs. Similarly, Israel started to deploy new Merkava tanks, though the bulk of its tank force was made up from older American M-60s. Iraq was able to procure modern weapons from the Soviet Union and France, such as MIG-23s and Mirage F1s, though none of its major imported weapons systems featured the state-of-the-art sophistication of Israeli F-15s and F-16s or Merkavas.[72] This advantage may have been vital for Israeli success in Osiraq, particularly due to the F-16s being equipped with a modern computerized bombing system which allowed the Israeli air force destroying the reactor with great accuracy and little collateral damage.[73]

This advantage would have been of little use unless Israel's intelligence had managed to collect sufficient information about the target. Apparently, Israel's intelligence services managed to penetrate the Iraqi nuclear program. Claire claims that an Iraqi official was compromised while in France and transferred the plans of the Al Tuwaitha nuclear complex into Mossad's hands.[74] Other sources suggest that some French and Italian workers were recruited by Israel's intelligence between September 30, when the Iranian air raid made most foreign technicians working at Al Tuwaitha leave the country, and April 1981 when they returned.[75] Either way, Israel's intelligence was able to get vital information about Osiraq, enough to facilitate the raid. Reportedly they discovered that the air defense troops assigned to the complex regularly went to dinner before sunset; the Israeli raid was then scheduled for this moment.[76] Israel also gained an almost unrestricted access to U.S. KH-11 reconnaissance satellite imagery, which was denied even to U.S. closest allies in NATO.[77]

Intelligence was essential in facilitating the strike, yet political motivations were the cornerstone in the Israeli decision to strike. The conflict with Iraq was certainly perceived through the lens of a broader Arab-Israeli conflict. At the end of the day, it was only some eight years since the Yom Kippur War that was widely regarded as disastrous in Israel. Iraq had participated in the 1948, 1967, and 1973 wars and as Perlmutter, Hande, and Bar-Joseph observe, "unlike any other Arab state directly at war with Israel, Iraq consistently and stubbornly refused even to consider the conclusion of a ceasefire or armistice agreement with Israel."[78] Furthermore, Bagdad openly supported various radical Arab terrorists.[79] The Arab-Israel conflict was unsurprisingly the most central to the Tel Aviv decision makers. Under such circumstances, the Iraqi reactor was viewed as a threat to the existence of Israel by many.[80] In fact, a conflict with nuclear weapons was of greatest concern to Israeli leaders and reminded them of the mass atrocities against the Jews during the Holocaust.[81]

This threat perception goes hand-in-hand with Israeli perceptions of Saddam's regime as highly irrational and risk-prone. Begin considered Iraq "the bloodiest and most irresponsible of all Arab regimes, with the exception of Kaddafi in Libya."[82] This view is unsurprising given the fact that Saddam publicly predicted "that the Arab nation would survive even if Israel killed 40 million Arabs."[83] Apart from Saddam's wild rhetoric, his domestic behavior also contributed to the fears of Israeli leaders. Saddam legally succeeded Iraqi President Hassan al-Bakr after his resignation in what Neeman describes to be the first normal succession (not a violent coup) in the history of the Iraqi republic, yet after this seemingly peaceful transfer of power, Saddam purged the military of 30 generals and 500 other officers.[84] The fact that Iraq invaded Iran only a few years after the two countries signed the Algiers Agreement that was intended to settle mutual conflicts only added to Israel's perception of the irresponsibility and aggressiveness of the Iraqi regime. Emerging reports of Iraqi usage of chemical weapons in this war that started to appear in 1980 fitted only too well into this perception.

For Israel there was little reason to believe that Saddam was an irresolute enemy that would easily capitulate to the threat of Israel's military might. In fact, the possibility that Saddam would retaliate in kind was discussed by Israeli decision makers during the critical cabinet meeting in October 1980.[85] It was also argued that the destruction of Osiraq would only strengthen Saddam's commitment to get nuclear weapons, something that actually happened after the strike, suggesting that Saddam's resolve was certainly not perceived as a weak one.[86] The Israelis were not unaware. In an attempt to undermine such further efforts at nuclear armament, Begin and Sharon declared what would be known as the Begin Doctrine, stating that Israel will not allow its enemies to acquire nuclear weapons.[87]

The views of Saddam's personality mated with a strong history of hostility between Iraq and Israel and the almost nonexistent institutionalization of mutual relations. As has already been mentioned, Iraq had participated in major Arab wars against Israel. While Iraq's contribution paled in comparison to Egypt and Syria, its role was hardly negligible. In the last war (1973) the presence of Iraq's expeditionary force of some 30,000 men, 500 tanks and 700 APCs initially caught even the IDF by surprise.[88] Subsequently, when Israeli-Egypt relations improved and in the wake of Camp David, Saddam claimed the leadership of the Arab anti-Israeli camp. Naturally Israel did not hesitate to repay the Iraqis for the favor. In the 1970s, Tel Aviv supported a Kurdish rebellion against Bagdad and despite the harsh anti-Israel rhetoric of the Iranian revolutionary leadership, Israel also gave its support to Iran during the Iran–Iraq war.[89]

Not only were relations between Israel and Iraq traditionally hostile, but they also remarkably lacked formal and informal communication channels. While Israel's equally hostile Arab neighbors often chose to establish at least some low-level working contacts and make formal and informal ceasefire agreements with their Israeli counterparts, the Iraqis simply withdrew from the battlefield.[90] In the absence of direct communication channels, Israel had to turn to public statements to warn the Iraqis indirectly about its positions.[91] It is no surprise that in its explanation of the attack against Osiraq, the Israeli government recalled "Iraq's declared maintenance of a state of war with Israel and its persistent denial of Israel's right to exist."[92] Under such circumstances, striking Osiraq seemed inevitable to many leaders in Israel.

This was certainly the case with Begin. To the Israeli Prime Minister, June 1981 appeared as the last resort moment. First, Osiraq was soon to become hot and Israeli intelligence had warned Begin that striking the reactor after it became operational would cause extensive civilian casualities.[93] Thousands of civilians killed by radioactivity in Bagdad – only a few kilometers north to Al Tuwaitha – was not something Begin wanted to accept. Second, the Israeli elections were scheduled for Fall and Likud was lagging behind Peres's Labour Party. Begin had serious doubts that, if he lost the elections, a new government would let Osiraq operate instead of destroying it.[94] The last resort feeling was not universal, though it prevailed among the supporters of the strike. The opponents, including Peres, argued that the time was not ripe for an attack as Iraq would need many years to get the bomb, offering Israel time for alternative non-military solutions.[95]

In 1981, with Israel's precarious position in the volatile Middle East, international factors were anything but excluded from the discussions among Israeli decision makers. Third-party deterrence was probably the least disturbing of the bunch. Not many countries were in a position to endanger Israel militarily. Israeli decision makers discussed the possibility that the strike would unite Arabs against Israel again, but Arabs had tried several times and had been unable to defeat Israel's military might. Furthermore, Iraq was able to muster substantial rhetorical support due to its position in the bloc of third-world countries, but lacked real allies. The most fearsome plausible third-party deterrer was the USSR.[96] But Iraq was not a typical Soviet client state.[97] At the end of the day, Saddam had expelled Soviet advisors from Iraq in 1972.[98] The Soviets, known as anti-proliferation hawks, continued selling military hardware to Iraq, but were not enthusiastic about Bagdad's nuclear program.[99] Besides, Moscow had enough problems with its economy, and the war in Afghanistan, to be eager about getting involved in Iraq's defense. Similarly, the French were interested in Iraq's oil and money, but not in defending its nuclear ambitions. The relatively mild reaction of Paris when it learned about the Israeli destruction of the French-made reactor and death of a French citizen there is revealing enough.

What Tel Aviv feared more than a possible hostile action by any third party was the impact of the strike on its unwritten alliance with the United States. During cabinet meetings, Mossad head Yitzhak Hofi profoundly expressed his doubts about the strike's impact on U.S.–Israeli relations, arguing that a possible rift in the alliance would be more damaging to Israel's security than Iraq's nuclear reactor.[100] His expectation was not unwarranted. While President Reagan was deeply pro-Israel, the position of his administration was not unanimous.[101] In Israel, the concerns about U.S. reaction were serious even among the supporters of the strike. While the United States was not consulted in advance, Begin promptly sent an explanatory note to Reagan after the strike.[102] The U.S. was disturbed anyway. Washington cooperated in drafting a UN resolution condemning Israel. The sale of further F-16s was suspended.[103] Yet America's anger was only partial and short-lived. The United States blocked any efforts to impose sanctions on Israel in the UN and effectively killed the move to expel Israel from the IAEA.[104] By September the sale of F-16s was quietly resumed.[105]

Considering the fact that in the 1980s most countries condemned Israel anyway, and apart from U.S. support there was not much to lose for Israel, the Tel Aviv government showed remarkable sensitivity to international action legitimacy. Hofi, Saguy, and Yadin cite the violation of Iraq's sovereignty among the reasons why they opposed the strike.[106] To enhance action legitimacy, the Israelis apparently attacked before the reactor was operational to prevent killing thousands of civilians by fallout, despite the fact that the destruction of a hot reactor would make rebuilding the facility more demanding. Israel also selected a Sunday date, expecting that French and Italian workers would not be on site, in order to reduce international criticism.[107]

While the evidence is incomplete, it also appears that the second normative aspect in the list may have played a role in Israeli considerations. Over the course of years, a nuclear taboo had developed in Israel that prohibited the use of nuclear weapons unless the state's existence was directly at stake.[108] Certainly there was no need to use nuclear weapons against Osiraq, as its destruction turned to be well within the capacity of conventional bombing. Yet it seems that the decision to strike Osiraq before it went hot was made on both normative and rationalist grounds. Begin did not only want to mitigate foreign opposition against the attack but also to limit collateral damage, especially on "the city of Baghdad and its innocent citizens," the latter on moral grounds.[109]

The action's legitimacy with a wider domestic audience was not a problem. Nothing suggests that the issue of domestic action legitimacy was even raised during cabinet meetings. In fact, it was even speculated after the strike that among Begin's motivations had been an attempt to increase his fading popularity. Regardless of the validity of such speculations, the raid enhanced Begin's popularity, with the nation allowing him to win the election on June 30, 1981, though with the smallest possible margin.[110] Yet, in many respects, the domestic constraints turned out to be at least equally limiting for Begin as international ones.

The Israeli population could have been sympathetic to the raid, but the Prime Minister had to fight a fierce battle with his senior officials and cabinet colleagues. In fact, national security professionals and cabinet members were deeply divided. Both intelligence chiefs, Hofi and Saguy, stood firmly against the strike. But their services were divided. Deputy Director of Mossad Nahum Admoni as well as the head of AMAN's Research Department Avi Ya'ari openly supported the strike during critical meetings against their bosses.[111] Minister of Defense Ezer Weitzman even resigned in protest to the strike.[112] Begin's Deputy Prime Minister Yadin also stood firmly among the opponents and reportedly threatened to resign, though he probably dropped his objections after he was presented with raw intelligence on March 15, 1981.[113]

Such intragovernmental deliberations were certainly related to another factor in the list, namely regime type. Israel was a vibrant democracy with open competition among a number of parties. Coalition governments usually forced Israeli prime ministers to make concessions to smaller partners. The militaristic character of Israel stems from the country's dangerous geographical location and violent history with its neighbors. Most of the country's policy makers were former military men and almost all of its citizens had served a long compulsory service: Begin had experience from a terrorist guerilla against the British; Yadin, a former general, had served as IDF Chief of Staff from 1949 to 1952; similarly, Weitzman was an experienced combat pilot and former deputy Chief of Staff of IDF; the raid's staunched supporter, experienced field commander and decorated war hero Ariel Sharon was widely regarded as one of Israel's best military strategists. Despite the country's characteristics, the constitutional balance that characterizes civil–military relations in Western democracies was generally maintained in Israel, though the IDF had a somewhat disproportional influence in shaping Israel's military strategy.[114]

Knowing the result, a current observer will not be surprised that none of the four possible deterrence threats Iraqis could muster against Israel managed to influence the decision of Begin's cabinet. Some concerns were apparently raised about the possibility of Iraqi retaliation on Israel's nuclear complex in Dimona, likely by conventional weapons, not by any bomb made out of diverted plutonium. Reportedly precautions were taken to tackle such a challenge. No extensive discussions are known to have been made either about Saddam's possible nuclear retaliation with diverted plutonium, or about nuclear denial.

With respect to conventional retaliation, it was well known that Saddam's hands were tied. His supposed ability to deploy five divisions against Israel could be fearsome, yet by 1981 all available Iraqi units were badly needed to battle the resurgent Iranians. Iraq could retaliate with its air force or even SCUD missiles. The first option would have been difficult against the advanced Israeli air defense yet, surprisingly, there is also no evidence that the Israelis discussed the possibility of retaliation with SCUDs. While Saddam was reportedly deterred from using SCUDs with chemical weapons against Israel, it is unclear why the SCUD threat was not at least discussed by the Israeli government.

More difficult to overcome was Iraq's conventional denial. After the largely symbolic Iranian air raid, Osiraq's air defense was increased, complicating further the already difficult Israeli mission. But Israeli intelligence helped to relieve some of the fears by discovering the loophole in the Iraqi air defense, namely the timing of the troops' dinner. It is unlikely that the unavailability of this information would have been prohibitive considering the strong feeling Begin had against the reactor. It was estimated that the strike package would suffer substantial losses and the Israelis proceeded anyway.

The empirical results summarized in Table 4.1 allow several interesting observations. Most notably, deterrence failure appeared in both conventional and nuclear dimensions. This is in marked difference with previous cases, where nuclear deterrence failed, but conventional deterrence held. In contrast to them, Iraq cannot credibly threaten robust conventional retaliation against Israel as its forces were tied up in a bloody war of attrition with Iran. Thus despite the fears over alliance relations with the United States, over international action legitimacy, and over influential opposition inside the government, the last resort feelings of Prime Minister Begin and other hawks prevailed, leading to the decision to launch the best known preventive counter proliferation strike.

Table 4.1 Israel–Iraq, 1977–1981

Concept	Description	Values
The deterrer's nuclear arsenal	Key physical attributes of respective nuclear complexes; weapon types; numbers; delivery vehicles; command, control, and communication systems	0–1 (weapon-grade HEU from France; reactor nearing completion)
Nuclear asymmetry	Situation when a substantially smaller (quantitatively) and less sophisticated nuclear force (qualitatively) faces a qualitatively and quantitatively larger nuclear force	Strong asymmetry (quantitative superiority by a factor of at least 100 and comparable quantitative edge)
Second-strike criterion	Nuclear posture that has the ability to survive enemy attack, make and communicate decision to retaliate, overcome enemy's active defense, and destroy a valuable target despite its passive defense	None (if HEU was diverted for the bomb it would have to be delivered by Iraq's poor air force against Israel's sophisticated air defense)
General conventional preponderance	Situation when one side's armed forces are in general substantially stronger in terms of numbers, technology, training, and employment strategies	Challenger (the deterrer's forces tied down in war with Iran)
Theater conventional preponderance	Situation when one side's armed forces are substantially stronger in terms of numbers, technology, training, and employment strategies on a theater where targets valuable to the challenger can be found	Challenger (with regional character of dyad theater and general overlaps)
The challenger's technological advantage	Significant advantage of the challenger's major weapons systems that would be employed in case of conflict in terms of state-of-the-art sophistication over the deterrer's weapons systems	Strong (significant edge in sophistication of critical weapons platforms)

Availability of information	The challenger's knowledge about targets' location and defensive systems, which is established from sources that the challenger deems credible	Strong (Israeli intelligence penetrated Iraqi nuclear program and also got satellite pictures from the U.S.)
Centrality of conflict	Absolute importance of the conflict dyad between the challenger and the deterrer from the challenger's subjective perspective and its relative importance to other existing conflict dyads where the challenger is a party	Yes (perceived through the lens of the Arab-Israeli conflict; Iraq appeared as one of the most hostile Arab nations)
Perceived resolve of the deterrer	The challenger's perception of the deterrer's commitment to fight should the deterrence fail	Strong (Saddam believed would try to retaliate, actions taken to mitigate it)
Institutionalization of mutual relations	Degree of shared expectations about the requirements of stable deterrence and the existence of proven formal or informal communication channels	Limited (even working-level relations were almost nonexistent)
History of hostility	Track record of conflict in the dyad that shapes the understanding of other side's intentions	Strong (Iraq participated in three wars against Israel, never signed cease fire)
Last resort	Situation when the challenger sees only the options to strike, or to live up to the development he tries to prevent	Strong (reactor about to be operational; elections scheduled; last-resortness not shared by opponents of strike)
Perception of the deterrer's rationality	The challenger's perception of the deterrer's rationality, particularly whether the challenger believes it possible to live with the nuclear-armed deterrer in the long term	No (Saddam's domestic behavior, anti-Israeli rhetoric, use of chemical weapons, invasion of Iran)

continued

Table 4.1 Continued

Concept	Description	Values
Third-party deterrence	Threat of military involvement into the original conflict by a third party, most likely the allies of the original deterrer, that decisively influenced the challenger's decision	Limited (Iraq had no real allies, or they were not able to endanger Israel militarily)
Alliance politics	Sensitivity of the challenger to possible impact of his action on the relations with his allies	Strong (fears over future of alliance with the U.S. expressed vocally within the government)
International action legitimacy	The challenger's sensitivity to the international normative expectation of non-intervention	Strong (concerns about legitimacy raised; decision made to strike before reactor was hot to limit international anger)
Nuclear taboo	Normative prohibition on the use of nuclear weapons	At least emerging (normative sensitivity to civilian causalities in Baghdad, nuclear option not even discussed)
Domestic action legitimacy	Level of support for military solution among the challenger's population	Strong (after-attack polls showed remarkable approval rate)
Opposition among influential decision makers	Negative view on military solution by influential part of the challenger's government	Strong (including deputy prime minister, minister of defense and heads of intelligence services)
Regime type	The challenger's position on the democracy-nondemocracy axis and on the militarized-nonmilitarized axis	Democracy, civilian control of armed forces (with strong military background of majority of decision makers)

Nuclear retaliation	Threat that the deterrer will use its nuclear weapons against targets valuable to the challenger, except of targets that are directly related to pursuit of the challenger's objectives	Limited (no concerns about Iraq's nuclear retaliation raised in the government; no preparatory measures taken to mitigate such threat)
Nuclear denial	Threat that the deterrer will use its nuclear weapons against targets that are directly related to pursuit of the challenger's objectives in order to prevent him from attaining the objectives, or in order to make it unacceptably costly	Limited (no concerns about Iraq's nuclear denial raised in the government; no preparatory measures taken to mitigate such threat)
Conventional denial	Threat that the deterrer will use its conventional weapons against targets that are directly related to pursuit of the challenger's objectives in order to prevent him from attaining the objectives, or in order to make it unacceptably costly	Limited (concerns about Osiraq's air defense raised; losses expected; not considered prohibitive)
Conventional retaliation	Threat that the deterrer will use its conventional weapons against targets valuable to the challenger, except of targets that are directly related to pursuit of the challenger's objectives	Limited (concerns about retaliation against Dimona raised; defense prepared; most Iraqi forces tied against Iran)

Notes

1 Yuval Neeman, "The Franco-Iraqi Project," *Bulletin of Atomic Scientists*, 37/7 (August/September 1981) 8.
2 Målfrid Braut-Hegghammer, "Revisiting Osirak: Preventive Attack and Nuclear Proliferation Risk," *International Security*, 36/1 (Summer 2011) 105.
3 Jeremy Tamsett, "The Israeli Bombing of Osiraq Reconsidered: Successful Counterproliferation?" *The Nonproliferation Review*, 11/3 (2004) 73.
4 Rodger W. Claire, *Raid on the Sun: Inside Israel's Secret Campaign that Denied Saddam the Bomb* (New York: Broadway Book, 2004) 44.
5 Hal Brands and David Palkki, "Saddam, Israel, and the Bomb: Nuclear Alarmism Justified?," *International Security*, 36/1 (Summer 2011) 146.
6 Braut-Hegghammer, "Revisiting Osirak," 105–110.
7 Claire, *Raid on the Sun*, 49–51.
8 Amos Perlmutter, Michael Handel, and Uri Bar-Joseph, *Two Minutes over Baghdad* (London: Frank Class Publishers, 2005) 38.
9 Shai Feldman, "The Bombing of Osiraq – Revisited," *International* Security, 7/2 (Autumn 1982) 115–116.
10 Perlmutter, Handel, and Bar-Joseph, *Two Minutes over Baghdad¸* 47–48; Claire, *Raid on the Sun*, 59.
11 Braut-Hegghammer, "Revisiting Osirak," 112.
12 The technical details are mostly drawn from Braut-Hegghammer, "Revisiting Osirak," 110–114; Dan Reiter, "Preventive Attack against Nuclear Programs and the 'Success' at Osiraq," *The Nonproliferation Review*, 12/2 (July 2005) 358–359.
13 Brands and Palkki, "Saddam, Israel, and the Bomb," 150–155.
14 Rachel Bzostek, *Why Not Preempt?* (Abingdon: Ashagate, 2008) 148.
15 Claire, *Raid on the Sun*, 52.
16 Perlmutter, Handel, and Bar-Joseph, *Two Minutes over Baghdad¸* 43–44
17 Feldman, "The Bombing of Osiraq – Revisited," 117.
18 Braut-Hegghammer, "Revisiting Osirak," 111.
19 Feldman, "The Bombing of Osiraq – Revisited," 117.
20 Braut-Hegghammer, "Revisiting Osirak," 111; Feldman, "The Bombing of Osiraq – Revisited," 117.
21 Perlmutter, Handel, and Bar-Joseph, *Two Minutes over Baghdad¸* xxxvii.
22 Feldman, "The Bombing of Osiraq – Revisited," 115.
23 Neeman, "The Franco-Iraqi Project," 9.
24 See Avner Cohen, *Worst-Kept Secret: Israel's Bargain with the Bomb* (New York: Columbia University Press, 2010).
25 Claire, *Raid on the Sun*, 53.
26 Perlmutter, Handel, and Bar-Joseph, *Two Minutes over Baghdad¸* xxxiv.
27 Bzostek, *Why Not Preempt?*, 149.
28 Perlmutter, Handel, and Bar-Joseph, *Two Minutes over Baghdad¸* 49; Claire, *Raid on the Sun*, 57.
29 Perlmutter, Handel, and Bar-Joseph, *Two Minutes over Baghdad*, 51.
30 Perlmutter, Handel, and Bar-Joseph, *Two Minutes over Baghdad¸* 56.
31 An exception was the mysterious death of Marie-Claude Magal, a witness in the case of El Meshad's murder and reportedly a prostitute occasionally hired by Mossad. Claire, *Raid on the Sun*, 68.
32 Reiter, "Preventive Attack against Nuclear Programs and the 'Success' at Osiraq," 357.

33 Claire, *Raid on the Sun*, 62.
34 The accounts are from Braut-Hegghammer, "Revisiting Osirak," 108–117; the author draws extensively from Jaffar's and Shahristani's accounts.
35 Brands and Palkki, "Saddam, Israel, and the Bomb," 133.
36 Brands and Palkki, "Saddam, Israel, and the Bomb," 133–166.
37 See Reiter, "Preventive Attack against Nuclear Programs and the 'Success' at Osiraq," 357; Braut-Hegghammer, "Revisiting Osirak," 108–109.
38 Perlmutter, Handel, and Bar-Joseph, *Two Minutes over Baghdad*, 60.
39 Feldman, "The Bombing of Osiraq – Revisited," 121; Perlmutter, Handel, and Bar-Joseph, *Two Minutes over Baghdad*, 60.
40 Lyle J. Goldstein, *Preventive Attack and Weapons of Mass Destruction: A Comparative Historical Analysis* (Stanford: Stanford University Press, 2006) 118; also see Perlmutter, Handel, and Bar-Joseph, *Two Minutes over Baghdad*, 67–83.
41 Here, hot refueling essentially means that the fuel is pumped into an aircraft which has already done all prestart procedures and has its engines running. It is intended to replace the fuel that was burned in the prestart period.
42 Claire, *Raid on the Sun*, 124.
43 Whitney Raas and Austin Long, "Osirak Redux?: Accessing Israeli Capabilities to Destroy Iranian Nuclear Facilities," *International Security*, 31/4 (Spring 2007) 10.
44 Perlmutter, Handel, and Bar-Joseph, *Two Minutes over Baghdad*, xl–xli; Claire, *Raid on the Sun*, 95; Bzostek, *Why Not Preempt?*, 154–157.
45 Goldstein, *Preventive Attack and Weapons of Mass Destruction*, 119.
46 Perlmutter, Handel, and Bar-Joseph, *Two Minutes over Baghdad*, xlii.
47 Perlmutter, Handel, and Bar-Joseph, *Two Minutes over Baghdad*, xliv.
48 Claire, *Raid on the Sun*, 129.
49 The text is available in Perlmutter, Handel, and Bar-Joseph, *Two Minutes over Baghdad*, 63.
50 Perlmutter, Handel, and Bar-Joseph, *Two Minutes over Baghdad*, 147.
51 Perlmutter, Handel, and Bar-Joseph, *Two Minutes over Baghdad*, 63; Claire, *Raid on the Sun*, 131.
52 Perlmutter, Handel, and Bar-Joseph, *Two Minutes over Baghdad*, 151.
53 See Paul Lewis, "Mitterrand Plans to Assure Arabs France Will Continue Arms Sales," *New York Times*, 27 May, 1981: A.7.
54 See Goldstein, *Preventive Attack and Weapons of Mass Destruction*, 117; Perlmutter, Handel, and Bar-Joseph, *Two Minutes over Baghdad*, 127; Braut-Hegghammer, "Revisiting Osirak," 111.
55 Tamsett, "The Israeli Bombing of Osiraq Reconsidered," 70; Claire, *Raid on the Sun*, 94.
56 Braut-Hegghammer, "Revisiting Osirak," 111.
57 Tamset claims that 12 kg is only half of what is typically required for a basic weapons design, but Braut-Hegghammer and Aroson and Borsh argue that 12 kg may suffice; see Tamsett, "The Israeli Bombing of Osiraq Reconsidered," 74; Braut-Hegghammer, "Revisiting Osirak," 112; Shlomo Aroson and Oded Brosh, *The Politics and Strategy of Nuclear Weapons in the Middle East: Opacity, Theory, and Reality, 1960–1991 – An Israeli Perspective* (Albany: State University of New York Press, 1992) 174.
58 Tom Sauer, "A Second Nuclear Revolution: From Nuclear Primacy to Post-Existential Deterrence," *Journal of Strategic Studies*, 32/5 (October 2009) 745–767.

59 Perlmutter, Handel, and Bar-Joseph, *Two Minutes over Baghdad*, 32.
60 Feldman, "The Bombing of Osiraq – Revisited," 124.
61 Brands and Palkki, "Saddam, Israel, and the Bomb," 157–161.
62 Perlmutter, Handel, and Bar-Joseph, *Two Minutes over Baghdad*, 35.
63 David Rodman, *Sword and Shield of Zion: The Israel Air Force in the Arab-Israeli Conflict, 1948–2012* (Brighton: Sussex Academic Press, 2013) 27–39.
64 Claire, *Raid on the Sun*, 53; Perlmutter, Handel, and Bar-Joseph, *Two Minutes over Baghdad*, 35.
65 Anthony H. Cordesman, *The Military Balance in the Middle* East (Westport: Praeger, 2004) 155.
66 Perlmutter, Handel, and Bar-Joseph, *Two Minutes over Baghdad*, 86.
67 Avner Yaniv, "Israeli National Security in the 1980s: The Crisis of Overload," in Gregory S. Maher (ed.), *Israel After Begin* (Albany: State University of New York Press, 1990) 97.
68 Williamson Murray and Kevin Woods, *The Iran–Iraq War: A Military and Strategic History* (Cambridge: Cambridge University Press, 2014) 85–171.
69 Brands and Palkki, "Saddam, Israel, and the Bomb."
70 Brands and Palkki, "Saddam, Israel, and the Bomb," 143.
71 Brands and Palkki, "Saddam, Israel, and the Bomb," 151.
72 Perlmutter, Handel, and Bar-Joseph, *Two Minutes over Baghdad*, 172–173.
73 Claire, *Raid on the Sun*, 75; Raas and Long, "Osirak Redux?," 16.
74 Claire, *Raid on the Sun*, 63.
75 Perlmutter, Handel, and Bar-Joseph, *Two Minutes over Baghdad*, 81.
76 Claire, *Raid on the Sun*, 180.
77 John J. Mearsheimer and Stephen M. Walt, *The Israel Lobby and U.S. Foreign Policy* (New York: Farrar, Straus and Giroux, 2007) 254.
78 Perlmutter, Handel, and Bar-Joseph, *Two Minutes over Baghdad*, xix.
79 Perlmutter, Handel, and Bar-Joseph, *Two Minutes over Baghdad*, xix.
80 Bzostek, *Why Not Preempt?*, 157.
81 Goldstein, *Preventive Attack and Weapons of Mass Destruction*, 118–119.
82 Brands and Palkki, "Saddam, Israel, and the Bomb," 155.
83 Brands and Palkki, "Saddam, Israel, and the Bomb," 165.
84 Neeman, "The Franco-Iraqi Project," 8.
85 Perlmutter, Handel, and Bar-Joseph, *Two Minutes over Baghdad*, xli.
86 See Braut-Hegghammer, "Revisiting Osirak."
87 Aroson and Brosh, *The Politics and Strategy of Nuclear Weapons in the Middle East*, 167–184.
88 Ian J. Bickerton, *Arab-Israeli Conflict: A History* (London: Reaktion Books, 2009) 133–134.
89 Brands and Palkki, "Saddam, Israel, and the Bomb," fn. 22; on Israeli support of the Iranians, see Trita Parsi, *Treacherous Alliance: The Secret Dealings of Israel, Iran, and the United States* (New Haven: Yale University Press, 2007).
90 Perlmutter, Handel, and Bar-Joseph, *Two Minutes over Baghdad*, xix.
91 See Perlmutter, Handel, and Bar-Joseph, *Two Minutes over Baghdad*, 54–58.
92 Bzostek, *Why Not Preempt?*, 159.
93 Perlmutter, Handel, and Bar-Joseph, *Two Minutes over Baghdad*, xxxvi.
94 Claire, *Raid on the Sun*, 96.
95 Bzostek, *Why Not Preempt?*, 157.
96 Aroson and Brosh, *The Politics and Strategy of Nuclear Weapons in the Middle East*, 175.
97 Goldstein, *Preventive Attack and Weapons of Mass Destruction*, 122.

98 Claire, *Raid on the Sun*, 47.

99 In fact, the USSR even imposed an embargo on military sales to Iraq on several occasions to punish Baghdad for its treatment of Iraqi communists, see Keith Krause, "Military Statecraft: Power and Influence in Soviet and American Arms Transfer Relationship," *International Studies Quarterly*, 35/3 (September 1991) 330.

100 Claire, *Raid on the Sun*, 54.

101 Claire, *Raid on the Sun*, 186.

102 Perlmutter, Handel, and Bar-Joseph, *Two Minutes over Baghdad*, 141.

103 Feldman, "The Bombing of Osiraq – Revisited," 128–131.

104 Feldman, "The Bombing of Osiraq – Revisited," 136.

105 Claire, *Raid on the Sun*, 9.

106 Claire, *Raid on the Sun*, 54.

107 Perlmutter, Handel, and Bar-Joseph, *Two Minutes over Baghdad*, 78.

108 Avner Cohen, "Nuclear Arms in Crisis under Secrecy: Israel and the Lessons of 1967 and 1973 Wars," in Peter R. Lavoy, Scott D. Sagan, and James J. Wirtz, *Planning the Unthinkable: How New Powers Will Use Nuclear, Biological, and Chemical Weapons* (Ithaca: Cornell University Press, 200) 105.

109 See statement of Israeli government in Perlmutter, Handel, and Bar-Joseph, *Two Minutes over Baghdad*, 120–121.

110 See Perlmutter, Handel, and Bar-Joseph, *Two Minutes over Baghdad*, 125–130.

111 See Perlmutter, Handel, and Bar-Joseph, *Two Minutes over Baghdad*, xli-xlii; Claire, *Raid on the Sun*, 26.

112 Claire, *Raid on the Sun*, 196.

113 Claire, *Raid on the Sun*, 196.

114 Yehuda Ben Meir, *Civil–Military Relations in Israel* (New York: Columbia University Press, 1995).

5   The United States and North Korea, 1992–1994

The end of the Cold War – one of the turning points in recent international history – separates the 1994 crisis on the Korean peninsula that almost culminated in a U.S. preventive strike against North Korea's nuclear installations from older cases. It allows enhancing the cross-case comparison with a more recent example, one not overshadowed by global bipolar competition. The limited availability of hard evidence from the inside debates of the Clinton administration, which often remains classified, is a necessary downside of the most recent case in this study. Unavoidably, certain details of the administration's planning remain unclear. Yet the seriousness of the crisis should not be underestimated. At the end of the day, there was the somewhat similar lesson of Osiraq, now seen in an entirely new light. By 1994, with the availability of the 1991 Gulf War experience, the American condemnation of the Osiraq destruction completely changed into a widespread appreciation of the attack. It is of little surprise that the possibility of launching a preventive strike against the Yongbyon nuclear complex was dubbed the Osiraq option.

The plot

North Korea had already embarked on the nuclear road in the 1950s. Driven by its *juche* ideology of self-reliance, Pyongyang initially attempted to get support for both a civilian and a military nuclear program from its Cold War allies. Yet even unscrupulous pressure, evasions and subterfuge did not help – Moscow and Beijing turned down the North Koreans' requests to help them build nuclear weapons. Pyongyang attempted to bypass the two socialist nuclear powers, approaching the Eastern Europeans directly for nuclear know-how, but Berlin, Prague, and Budapest followed the Soviet example.[1] The most Pyongyang secured from Moscow was the construction of a small research reactor in Yongbyon in 1965, and the proliferation-sensitive Soviets insisted on placing the reactor under the IAEA's inspection even though North Korea was not party to the NPT at the time.[2] Having been left without the desired help, North Korea started the construction of its indigenous 5 MW gas-graphite research reactor based on declassified British blueprints in 1979. The reactor went operational in January 1986. However, a small indigenous reactor could not decisively alter North Korea's energy shortages. Large Soviet light water reactors could. But the Soviets had a precondition: Pyongyang must first sign the NPT. In December 1985, the North Koreans conceded and signed. However, due to USSR's failing economy, the reactor deal never came to fruition. Pyongyang did little better in fulfilling its part of the deal. The safeguard protocol, which would allow the IAEA to monitor North Korea's nuclear sites, remained unratified until 1992.

The construction of an indigenous reactor in Yongbyon could not stay secret for long. The American CORONA satellite spotted its foundations in 1980.[3] More disturbing information was yet to come. In 1988, satellites detected a large facility under construction in Yongbyon which intelligence believed to be a reprocessing plant.[4] A year later, satellite photographs detected extensive tests of conventional explosives in a way corresponding to tests of explosives that would be used to detonate a nuclear weapon's plutonium core.[5] Officials in the Bush administration became concerned. Soon the ups and downs of nuclear diplomacy with North Korea started.

The initial development was promising. The North Koreans had complained for a long time about the presence of U.S. nuclear weapons on the Korean peninsula, conditioning cooperation here and there with their removal. The international situation was favorable. Reflecting the end of the Cold War, the Bush administration started a major review of nuclear weapons deployment. After some discussions, Korea was included in the plan. By the end of 1991, all U.S. nuclear weapons were removed.[6] Responding to this progress, both North and South Korea signed a Joint Declaration on the Denuclearization of the Korean Peninsula on December 31. As a reward, Pyongyang earned the cancelation of the 1992 annual *Team Spirit*, the major joint U.S.–South Korea military exercise which had regularly driven Pyongyang crazy and forced it to mobilize its troops.[7] This was not the only sweetener. On January 22, 1992, Undersecretary of State for Political Affairs Arnold Kanter met Secretary for International Affairs of the Korean Workers' Party Kim Yong-Sun for historical high-level talks in New York.[8] The negotiations did not make great progress over substance, yet being recognized as a negotiating partner with the Americans was understood as a great success for North Korea's diplomacy. By April 1992, it appeared that the positive development was going to prevail over occasional bumps.

In January, the DPRK signed the IAEA's safeguard agreement, promptly ratifying it on April 10. Subsequently, in May, Pyongyang provided the IAEA with the required data on its nuclear program, while the agency's director general Hans Blix paid a visit to Yongbyon.[9] Why exactly the North

Koreans agreed with the inspections is unclear. Likely Pyongyang underestimated the IAEA's technical abilities and desire to discover any discrepancies between reports and reality, something not surprising considering the IAEA's lax approach to inspections prior to the Gulf War. However the discoveries of an advanced nuclear weapons program in Iraq profoundly changed the IAEA that was now struggling to recover its lost credibility.[10] In any case, the July inspection in Yongbyon revealed that North Korea had separated plutonium in 1989, 1990, and 1991, not only in 1990 as it declared.[11] How much plutonium the North Koreans separated was unclear. Another round of inspections followed in September. The North Koreans were apparently trying to hide something. A previously two-floor building was now masked by bulldozed earth to appear as a single floor, something that could not avoid the attention of sophisticated U.S. satellite imagery. The inspectors were only allowed on to one of the two waste sites and only to take radiological measurements, not samples.[12] Under such conditions, the U.S. administration could not resist pressure from domestic hawks and the South Koreans to take a tougher stance.

The unsatisfactory inspections opened room for the first major crisis. Seoul, which was regularly haunted by incompatible fears that others would take the lead on the intra-Korean issue if its position towards the North was too weak, and that the conflict was dangerously close, pushed the United States into a resumption of Team Spirit in Fall 1992. The State Department opposed it vigorously, but the idea was received well in the Pentagon.[13] Other signals that the situation was about to deteriorate came from Vienna. There, on February 22, 1993, the IAEA's Board of Governors was presented with U.S. satellite photography confirming North Korea's cover-ups in Yongbyon. Under the burden of evidence, the Board had no other option but to call on North Korea to comply with its safeguard obligations and allow access into the two waste sites within one month.[14] Two weeks later, the Team Spirit started. Though it was downscaled to 70,000 South Korean and 50,000 American troops, North Korea was predictably outraged and ordered its troops to be battle-ready.[15]

The new governments in Washington and Seoul, both elected only in the last two months of 1992, were far from fully accustomed to their new jobs when the breaking news came from Pyongyang. On March 12, 1993, North Korea announced its withdrawal from the NPT. Such a step was legal under the treaty's Article X, yet surprising and alarming for both Washington and Seoul. Quickly, the Republic of Korea's forces went to a high state of alert, and an interagency meeting was hastily convened in Washington. It turned out that the North was not interested in further escalation. The intelligence confirmed that no unusual military moves of Pyongyang's army had been detected.[16] Instead of direct confrontation, what followed is well described as rounds of evasive maneuvers that mixed conciliatory gestures with unacceptable requests. Not only Pyongyang, but Washington, Seoul, and Vienna took their parts in this game.

Two weeks into its withdrawal declaration, North Korea "categorically" rejected IAEA's request to hold a special inspection of the waste sites, but offered "implementation of the safeguard agreement" at other installations.[17] Yet the IAEA jealously guarded its right of special inspections. On April 1, the agency declared the DPRK in violation of its safeguard agreement and reported the case to the UN Security Council.[18] The Clinton administration responded with a strategy of gradual escalation, expressing its support for the IAEA's demand for special inspections, and slowly started building a coalition for eventual sanctions.[19] In the meantime, the North Koreans started dropping hints that they were willing to cooperate and might conditionally suspend, or even reverse, their decision to leave the NPT.

The IAEA reluctantly accepted the North's offer to conduct limited inspections that would become known as "continuity of safeguard information." The term largely covered the maintenance of the agency's equipment in Yongbyon, including the necessary replacement of film and batteries in cameras monitoring the installations.[20] This need to maintain appeared pressing: the IAEA expected that the reactor in Yongbyon would be refueled in May and wanted to have an eye on the spent fuel that it expected to be extracted from the reactor. Yet it did not happen. In one of its conciliatory moves, Pyongyang decided to leave – at least for the moment – the spent fuel worth four to five bombs of plutonium in the reactor.[21] Washington also responded positively to Pyongyang's gestures and agreed to open negotiations. But first, to enhance pressure and its negotiating position, it secured the passing of UN Resolution 825 that urged Pyongyang not to proliferate.[22]

The first round of the U.S.–DPRK talks started in June. Reducing the talks from high level to policy level, Washington chose its proliferation expert and Assistant Secretary of State for Political-Military Affairs, Robert L. Galluci, as its lead negotiator. Surprisingly quickly, the meeting in New York brought positive results.[23] After the usual first-day exchange of opening broadsides and subtle threats, Galluci offered the North Koreans U.S. assurances that Washington would stick with its UN Charter obligations – such as to restrain from the use of force – but linked the issue to the DPRK's earlier statements to be ready to stay within the NPT. The North Koreans were ready to accept. Not only did the meeting culminate in the first ever joint statement between the United States and North Korea, but it tackled the threat of the DPRK's immediate withdrawal from the NPT. At the end of the negotiations, Pyongyang announced suspending its decision to leave the NPT while the talks with the United States continued.[24]

Another round of talks followed a month later in Geneva, but turned out to be less promising, unsurprising considering the scale of the issues that now comprised not only the immediate problem of NPT withdrawal but also

North Korea's nuclear future as well as past discrepancies. Despite the difficulties, the two sides managed to strike a fragile deal. The North Koreans agreed to open negotiations on inspections with the IAEA and on bilateral issues with Seoul, while "the United States would be prepared to support the introduction of light-water reactors in North Korea" even though only as a part of the final resolution of the crisis.[25] The deal's downside was apparent. Success was not only conditional on support from Washington and Pyongyang, but also on the agreement of Seoul and the IAEA. Soon the problems emerged at more than one side.

The IAEA insisted on its right to inspection, yet, for the North Koreans, disclosing their full nuclear past was unacceptable. The agency at least dispatched its inspectors to Yongbyon in August to take care of equipment maintenance and the limited inspections North Korea was ready to accept. There, the inspectors discovered a damaged seal at one of the hot cells. The agency concluded that no nuclear material could have been diverted, but the bitter taste remained.[26] In September, the battles between the IAEA and Pyongyang continued, whilst the cameras at Yongbyon were slowly running out of film. This time the IAEA was not willing to accept partial compliance. By September 28, the cameras in Yongbyon ran out of film. In a few weeks it was clear that this round of IAEA-DPRK talks would have a similar fate. The fate of the North Korea-South Korea talks was not dissimilar and, though Pyongyang unsurprisingly raised preconditions difficult to meet, it was not the only one to blame.[27]

As the talks fell apart, the Principals Committee gathered in the White House to discuss what to do. Signifying the importance of the issue, national security advisor Tony Lake, his deputy Sandy Berger, secretaries Christopher and Aspin, new chairman of the JSC Shalikashvili, Vice-President's national security advisor Leon Fuerth, and DCI R. James Woolsey were attending. While the principals in general endorsed the existing approach to the DPRK, this time the top national security leaders also reviewed the military situation and options. Lake raised the possibility of destroying Yongbyon, but Shalikashvili could not guarantee the destruction of the plutonium that was already extracted as Woolsey could not confirm its location. Shalikashvili also estimated that the attack would spark the DPRK's reaction across the 38th parallel.[28]

More promising news came from Pyongyang. On December 3, the North Koreans proposed to admit the IAEA to all the sites, though some activities would be prohibited at the reactor and the reprocessing plant. Washington's inclination was to accept the offer, but the IAEA was unenthusiastic. It took almost the whole month to strike a preliminary package deal, nicknamed "Super Tuesday." The deal envisioned four simultaneous steps: the IAEA would begin inspections; the North would resume talks with the South; Seoul would announce the cancellation of Team Spirit; and the U.S. would announce the date of the next meeting between Galluci and his North Korean counterparts.[29] The complicated plan did not take long to fall apart.

One unexpected blow came from the American side. After the Somalian fiasco, the Clinton administration was ready to give field commanders what they asked for. Already in December U.S. commander in Korea General Gary Luck asked for Patriot missile batteries to protect his troops. For the Americans it was an unproblematic defensive weapon. Yet for the North Koreans, the Patriots would cover U.S. entry points to the peninsula and allow the inflow of further reinforcements in a moment of crisis. Yet North Korea's only hope for a military victory was to prevent U.S. reinforcements from coming. No decision about deployment was made as of January when the media learned about and disclosed the possible Patriot deployment. But the North Koreans were unhappy anyway.[30]

Still, in February, the IAEA and the DPRK managed to agree on the inspections. A small downside was that both parties expected the other party to agree with their interpretation.[31] Unsurprisingly, the diverging interpretations hampered the inspections. The inspectors arrived to North Korea on March 1, but they soon learned that they would not be allowed to take all the samples they wanted.[32] Despite attempted negotiations, the differences were impossible to resolve. The IAEA decided to push its case. On March 15, it withdrew its inspectors and convened the Board of Governors meeting for March 21.[33] At the same time, the negotiations between North Korea and South Korea in Panmunjom failed when North Korean negotiators threatened their counterparts to "turn Seoul into the a sea of fire" and left.[34] The March 21 Board of Governors meeting only signified the death of Super Tuesday as it urged North Korea "immediately to allow the I.A.E.A. to complete all requested inspections activities and to comply fully with its safeguard agreement" and referred the case to the Security Council.[35] Pyongyang got itself into a state of dangerous isolation. Only Libya opposed, while China departed from its previous practice of voting against resolutions on the DPRK and abstained.[36]

In the following months the crisis heated to a new scale. The seriousness of the events led the United States in April to establish a special Senior Policy Steering Group on Korea. Chaired by Galluci, the group reported directly to the Principals Committee. On April 16, Galluci, together with Defense Secretary Perry, traveled to Seoul to consult with the ROK. While the two officials were in Korea, Kim Il-Sung gave an unusual public appearance and, surprisingly, he sounded conciliatory.[37] The aging leader may have been sincere but his opening was short-lived. Only a few days later, the first Patriot missiles arrived.[38] This time, the North Koreans responded strongly by informing the United States that they would refuel the reactor in Yongbyon. For many in Washington, refueling appeared as the red line: first, it would destroy IAEAs ability to uncover how much plutonium North Korea extracted before; and, second, the DPRK would be in a position to reprocess spent fuel and get plutonium for four to five bombs.

In mid-May, the actual process of fuel rod removal started. While it would take time to finish it – CIA and IAEA estimates varied between three to six months – the development was dangerous.[39] Once the removal started, the IAEA had no other chance than to accept the North's offer to monitor it despite limits imposed on inspectors. But the agency also felt compelled to refer a serious violation of safeguards to the UN. The case appeared to be getting into an escalating spiral. Washington canceled the next round of talks with the North Koreans and intensified its effort in building the coalition against Pyongyang.[40] On June 10, the IAEA initiated its own, largely symbolic sanction against North Korea, which responded three days later by announcing that it was leaving the agency and asserting that UN sanctions against Pyongyang would be considered a declaration of war.[41] North Korea's threat notwithstanding, the U.S. responded with a call for phased sanction. The issue was debated by the Principals Committee on June 14, when the draft resolution was agreed and promptly forwarded to Tokyo and Seoul for consent, which came a day later. But sanctions were not the only topic debated by the principals. As Wit, Poneman, and Galluci – two of them prominent members of Clinton's national security team dealing with North Korea – recall, "there was now a new focus, the Osiraq option."[42]

It was not the first time that the option to solve the problem of North Korea's nuclear program by preventive strike was discussed, yet now it was rising in prominence. Previously, the option had been discussed during the Pentagon's regular meetings on Korea.[43] In May, Secretary Perry gathered all the four-star generals and admirals in Pentagon to review war plans for a second Korean war. The fact that the results were then presented to the President and the Vice-President only signifies that the problem was not taken lightly.[44] By June, 14 principals were well-acquainted with the downsides of the Osiraq option. The most worrying was the likely escalation – the possibility that the DPRK's million-man strong army would move south. They decided to postpone the decision for later as it was agreed that the strike could take place only after U.S. reinforcements arrived in South Korea.[45]

But what would happen later was dangerously unclear. Washington was readying the coalition to impose the UN sanctions. South Korea and Japan agreed, Britain and France as known proliferation hawks could be counted on, Russia and China were hesitant, but appeared to be slowly moving to the pro-sanctions camp. Yet the North Koreans repeatedly declared that sanctions would equal war. Arguably, they were bluffing, but it was uncertain. To make matters worse, the military situation on the peninsula was precarious. Weak but armed to the teeth, North Korean's only hope for winning the war against the economically much stronger ROK, probably backed by rapidly incoming U.S. reinforcements, was to take the offensive and block the U.S. from coming. Acquainted with this strategy, the U.S. war plan for Korea changed in early 1990 to rely on its own offensive to prevent it.[46] Alongside war, Washington had other irons in the fire; most importantly, President Carter was en route to Pyongyang to meet directly with Kim Il Sung. Yet despite the approval and instruction from the U.S. government, Carter was traveling as a private citizen with limited options to offer to North Korea.

On June 16, President Clinton convened his council of war. Primarily, the top leaders should have decided the scale of reinforcements to be sent to South Korea. The options at sending 23,000 troops to prepare the logistics of the eventual arrival of 400,000 additional troops that the U.S. commander in Korea expected to need if war broke out. Other options were even more robust, including tens of aircraft and a second carrier group to be placed in the region.[47] Before the meeting, the President also reviewed the pros and cons of the Osiraq option.[48] Then, suddenly, during the second hour of the meeting, came the game changer. Former President Carter called from Pyongyang and informed the top national security officials of the acting administration, including the President, that he had struck a deal with Kim Il-Sung. North Korea would allow inspectors to stay and change its gas-graphite reactors for light-water ones. The United States would drop the sanctions motion. Jimmy Carter, an experienced politician hardened by numerous battles every U.S. president must fight not only with his foreign enemies but mostly with his own bureaucracy and U.S. media, well knew that he had offered the North Koreans more that he was allowed to. Thus to bind the administration's hands Carter promptly announced the deal on CNN.[49]

What exactly would have been decided at the June 16 meeting will remain unclear. Galluci, who was not only present at the meeting but actually answered Carter's call, argues that Clinton would have almost certainly approved the largest possible buildup, sending 50,000 soldiers to South Korea, as Perry and Shalikashvili had already decided to recommend. Then the sanctions would have been approved in the UN and, unless Pyongyang conceded, "the option Secretary Perry had decided not to recommend to the president – a military strike against Yongbyon – would have returned to the forefront."[50]

Unfolding the complexity

Two factors are particularly valuable in the North Korean case. First, it took place after the end of the Cold War; second, in November 1993, U.S. intelligence estimated that North Korea might actually be a nuclear power. The two are more than enough to deserve thorough the examination that will help penetrate the complexity of the case.

It was well known that a critical part of North Korea's nuclear weapons complex comprised a 5 MW gas-graphite moderated reactor and a plutonium separation plant in Yongbyon. More interesting was whether the North Koreans had used these facilities to produce a meaningful amount of plutonium and had made a bomb from it. Hard evidence was lacking. The North Koreans stubbornly refused to let inspectors find out how much plutonium had been extracted during Yongbyon's shutdowns in 1989, 1990, and 1991. The two plausible explanations for hiding this were completely different. Either North Korea had enough to build a bomb and perhaps had produced one, or it was far from there but did not want the IAEA confirm that the North was nuclear-weaponless. The U.S. intelligence community was far from certain. The CIA first reported by 1993 that in the late 1980s "North Korea ... ha[d] produced enough plutonium for at least one, and possibly two, nuclear weapons." This assessment was confirmed in further unclassified estimates and later turned into an assessment that the DPRK in fact had nuclear weapons by the early 1990s.[51] Certainly there were some indications. During the talks between Galluci and Kang in June 1993, the North Koreans dropped hints that they might have nuclear weapons.[52] South Korea's intelligence claimed that up to 22 kilograms of plutonium had been extracted.[53] Adding to the puzzle were recorded tests of conventional explosives for the weapon's trigger.[54] Nonetheless, the U.S. intelligence estimate's careful wording that "there was a better than even chance that North Korea had already produced one or two nuclear weapons" well demonstrated how uncertain the evidence was.[55]

Needless to say, with zero to two nuclear bombs, North Korea's arsenal was in a position of great asymmetry vis-à-vis the Americans. The fact that the United States had withdrawn nuclear weapons from Korea changed little to that. The U.S. strategic nuclear forces alone comprised of: 550 ICBMs, all of them MIRVed; 21 SLBMs with 440 Trident missiles, eight warheads each; and 190 strategic bombers equipped with a variety of 2,800 nuclear weapons.[56] All of the American weapons were highly sophisticated (older systems could have been retired with the end of the Cold War) and supported by an elaborate command and control system.

Apparently this would have left the DPRK's ability to strike back at a U.S. attack on a rather theoretical level. On the one side – assuming that one or two weapons really existed – the United States did not know where they were, making them possible survivors of an eventual first strike. Yet, as it has been noted, surviving does not equal having the ability to retaliate. Apparently, the delivery of the second strike was a weakness. In the Wall Street Journal, Mark Helprin claimed that Pyongyang's nuclear weapons were only deliverable by truck.[57] Likely, he was right. In the early 1990, the DPRK deployed a number of Soviet SCUD missiles and in fact tested its own larger variant called Nodong 1. Yet, beyond any reasonable doubt, the

DPRK did not master a technology that would allow tipping missiles with nuclear warheads by 1994. It is even dubious that the nuclear weapon could have been delivered by one of North Korea's aircraft. Even if it was, U.S. and ROK superiority in the air would likely down the attacker.

In general, conventional balance was not much more promising for the North Koreans. With the end of the Cold War, North Korea lost a large part of its invaluable support from the Soviet Union and China. It was an uncomfortable situation for Pyongyang: its economy, which outperformed its southern neighbor in the 1950s and 1960s, lost its pace in the 1970s and later went into a declining spiral. In the meantime, South Korea boomed. By 1993, a rather small country of some 23.9 million people, with a GDP of some U.S.$21 billion faced South Korea, which had grown to a country of 44.9 million people with a GDP of U.S.$300 billion.[58] Furthermore, South Korea's U.S. backing made the discussions about the North's ability to fight a prolonged war largely irrelevant. The United States had convincingly demonstrated their ability to deploy a large force to remote theatres just a few years ago in the Gulf War where they deployed 500,000 soldiers, 1,900 tanks, 930 artillery pieces, and 456 attack helicopters.[59] Some 35,000 U.S. troops were already stationed in South Korea. The ROK itself commanded a force of more than 600,000 soldiers, 1,800 tanks, and 4,400 artillery pieces, which was facing the DPRK's army of some 1.1 million men with 4,200 tanks, and 6,800 artillery pieces.[60] Notwithstanding the numerical superiority, North Korea was the significantly weaker party. Its weapons were generally less sophisticated than American equipment, half of its army was either soldier-workers or irregular paramilitary troops with insufficient training, and if the war broke out the United States and the ROK would enjoy command of the air.[61]

North Korea's best hope was in mustering local superiority for a decisive breakthrough and a blitzkrieg-like armored campaign that would prevent American reinforcements from coming. Bearing this in mind, the North Koreans transformed their army for an offensive posture in the 1980s. This change included significant mechanization, major reorganization of the army structure from separate divisions into corps-level formations, and pre-positioning of 70 percent of all combat forces into forward positions.[62] Still it was uncertain that a breakthrough could be achieved and sustained for a long time. The U.S. estimated that North Korea could not win the war even if it achieved strategic surprise.[63] The vastly superior training and readiness as well as air superiority of the allied forces would offset North Korea's advantage in numbers.[64] However, one significant strategic advantage of North Korea remained. The Seoul metropolitan area, the hub of the ROK's economy and population, was precariously close to the 38th parallel. It could be grabbed by an armored sweep. Even if such a move could be blocked by allied forces, which was uncertain, Seoul was within a range of thousands of the DPRK's heavy artillery tubes and the ROK lacked sufficient counter-battery radars to suppress their fire. This precarious situation was well known to the U.S. military. It was the reason why the principals decided in June that a strike could take place only after U.S. reinforcements arrived in Korea. Some were already deployed in the spring, including, most importantly, Patriot batteries and Apache helicopters, and the rapid insertion of anti-artillery systems was being prepared.[65] But more were needed to prevent North Korea's ability to achieve short-term local preponderance.

As it has been already mentioned, the U.S. enjoyed an unquestionably robust technological advantage. While in the early 1990s, the DPRK's armed forces deployed reasonably up-to-date Soviet systems, the Gulf War demonstrated that American technology was vastly superior. The DPRK's most modern weapons had arrived in 1984 when Moscow provided Pyongyang with SAM-5 missiles and MIG-29 and SU-25 fighters.[66] Yet all those systems were available to the Iraqis in 1991, with well-known results. Furthermore, in many respects, the DPRK's technology was in fact inferior to what was available to Iraq. For instance, North Korea had no T-72 tanks; the best it could deploy were local upgrades of the T-62. Of course, there were imperfections in the allies' superiority. Secretary Aspin complained about the insufficient ability of GBU-28 bombs to penetrate hardened targets while the North Koreans were masters at burying their assets underground.[67] Furthermore, the brunt of the ground offensive in its initial stage would be borne by ROK's forces that were less sophisticated then the Americans and who deployed a number of obsolescent weapons such as M-48 tanks along with indigenous modern K1s. Yet, the shortcomings notwithstanding, the new U.S. abilities in surveillance, air defense suppression, stealth technology, and precision-guided munition, as well as advanced ground-force technology like thermal sights, compound armor, and depleted uranium ammunition made the technological part of the hypothetical battle absolutely one-sided.[68]

More questionable, then, than the Americans' ability to destroy targets was their ability to locate them. Of course, U.S. satellites demonstrated impressive results, being able to detect North Korea's movements on an unprecedented scale. Yet satellites cannot detect everything. This was demonstrated by the failure of intelligence to detect the Iraqi nuclear program prior to the 1991 war and remained firm in everyone's minds by 1994. Human intelligence would be needed to supplement the satellites, but according to Sigal, U.S. intelligence lacked good sources with access to North Korea's top political authorities or its nuclear program. Interestingly, Sigal also argues that U.S. intelligence lacked extensive liaison with the ROK, which was better positioned to have agents in the northern part of the divided nation.[69] This corresponds with the information that the plutonium's location was unknown, which DCI Woolsey gave to the Principals Committee in November 1993.[70]

The imperfection of intelligence may well be attributed to the lower priority of the conflict prior to the 1993–1994 escalation periods. A number of other international conflicts such as Somalia, Rwanda, and Bosnia required the attention of an administration that was elected on the promise of focusing on domestic issues. The truth is that U.S. planners identified Korea as the most likely spot for hostilities in Asia since the end of the Vietnam War.[71] Yet, only slowly did the North Korean nuclear crisis move to the position of centrality. When the Clinton administration entered office, North Korea was not on the top of its nonproliferation agenda.[72] The administration's most important official tasked with managing the crisis, Robert Galluci, recalls that North Korea was not discussed with Secretary Warren Christopher until 1993.[73] Yet apparently things changed with the growing temperature on the peninsula. In November 1993, Secretary of Defense Aspin told reporters that North Korea was much more important than Bosnia and Somalia because "our interests are much greater in Asia."[74] From May 1994, North Korea got the President's upmost attention.[75]

Contrary to the centrality of the conflict, which markedly grew throughout early 1990s, three other perceptional factors under study – history of hostility, institutionalization of mutual relations, and perception of deterrer's resolve – remained virtually unchanged. All three were largely related to the formative moment in the U.S.–DPRK relations, the Korean War. For the Americans, this bloody conflict transformed and militarized the Cold War and cost them tens of thousands of lives.[76] By 1994, little had changed in mutual hostility. Even a decade later Jonathan Polack observes that the DPRK is "America's longest-running political-military adversary."[77] The Korean peninsula was still dominated by an "antagonistic Cold War atmosphere" well after its end everywhere else.[78] This historical perception was crowned by the mere fact that, 40 years after the armistice, North Korea was still led by Kim Il-Sung. As Wit, Poneman, and Galluci observe, "for the United States, holding a summit with the man who started the Korean War seemed out of the question."[79]

Such a perception of their mutual history unsurprisingly turned into little to no institutionalization of mutual relations. Only after Jimmy Carter's deal from Pyongyang culminated in the 1994 Agreed Framework did the United States and North Korea produce enough contact to find a language to move from their Hobbesian relationship.[80] Prior to the early 1990s, the two countries enjoyed only sporadic contact. Unsurprisingly, normal diplomatic relations did not exist. North Korea was not even a member of the United Nations before September 17, 1991.[81] As Leon Sigal observes, for the Americans "North Korea could have been on Mars."[82] The first high-level meeting between the two countries was only held in January 1992. There was no time to establish shared expectations about behavior. Galluci recalls that before his first meeting with the North Koreans a representative of the Joint Chiefs of Staff presented him with a book, *How Communists Negotiate*, which was written by a U.S. negotiator at the Korean War armistice talks, and "its tone reflected a cartoon-like image of the 'commies' typical of the 1950s."[83] The room for misunderstanding was enormous, but at least in one way did this history of hostility work to North Korea's advantage.

To the Americans, the North Koreans appeared stubborn enough, perhaps even crazy enough, to make their expected resolve very strong. Few would question Pyongyang's readiness to fight despite its meager prospects of defeating the United States and South Korea should the Korean War repeat. Decades of hostile relations showed that the North Koreans were unlikely to back down from confrontation. As Patrick Morgan observed, "[h]armful actions alone have been ineffective, mainly because the North seems more willing to suffer some escalation. It will not start another war but is willing to risk provoking one because its frustration with the status quo is greater."[84] Appearing resolute, tough, bellicose, and ready to take risk played a critical role in North Korea's deterrence strategy.[85] While this behavior more likely stemmed from North Koreans playing at rational irrationality than from them being really crazy, the Americans were uncertain.

That Pyongyang may be irrational cannot be ruled out. It probably even is, at least if Western normative and cultural standards serve as a benchmark to judge its behavior. The fact that in its international behavior a large part of the perceived irrationality may comprise of Schelling's rational irrationality was little comforting for the Americans. As one insider of the U.S. government in the 1994 crisis observed, "the basic assumption in the intelligence community and in the Defense was that North Koreans are liars, dug tunnels and can't be trusted."[86] Galluci's accounts are only slightly friendlier in his description of overall U.S. thinking about North Korea. He recalls that "[n]o one believed that Pyongyang could be trusted to carry out its agreements."[87] In this sense, it is largely surprising that the Americans, with the help of Jimmy Carter, managed to strike a deal with a country of dubious rationality and poor reputation for carrying out agreements instead of striking at the country itself.

The consensus in the administration, apparently, was that taking out Yongbyon was only a last resort option and that the last moment had not come yet. Yet if the North Koreans really were irrational and could not be trusted to fulfil agreements, striking it should have been the only option. Fortunately, uncertainty played into the hands of peace; there was still a chance that Pyongyang was rational enough to carry out what it promised. While the situation was never seen as the moment of last resort, it remains unclear how far away such a moment was. UN sanctions were very likely and massive U.S. reinforcements were almost certain to arrive should Jimmy Carter's intervention not work out. Since Pyongyang notoriously linked the sanctions to declarations of war and was expected to have a strong temptation to pre-empt if it expected U.S. reinforcements to arrive and inevitably end the slightest hopes the DPRK could have to prevail in the battle, the last resort perception could have changed rather quickly.

Of course, Pyongyang's saber-rattling was dangerous but hardly accidental. Largely, its roots go back to the DPRK's lack of allies to provide vital third party deterrence. During the Cold War, Pyongyang skillfully played China and the USSR, first to approve its adventurous invasion of the South, then to rescue it from the U.S.-led counteroffensive, and finally for decades to provide it with support which it rarely paid for. With the fading Cold War, the DPRK lost its value. Moscow was the first to let Pyongyang know. With its ill economy, it had few other options. In September 1990, Moscow normalized relations with Seoul and declared that it would not honor Soviet guarantees to the DPRK.[88] During the course of the crisis, Russia was often dissatisfied by its secondary role, but was generally supportive of U.S. policy.[89] China also had a mutual security pact with the DPRK and, at least on one occasion, subtly threatened the Americans that it might respond with a hostile reaction should the U.S. strike North Korea.[90] Yet the threat appeared largely non-credible. China forged an informal alliance with the U.S. in the 1970s, and normalized relations with Moscow, Tokyo, and Seoul over the course of late 1980s and early 1990s. None of that was advantageous to North Korea. That the relations between Beijing and Pyongyang were strained surfaced in the spring 1993 when the two countries' border guards fired on each other.[91] Of course, Beijing did not completely abandon Pyongyang. Both in the UN and in the IAEA, Chinese representatives were usually the most supportive of the DPRK of all relevant parties. Yet by mid-June, the U.S. largely successfully managed to get China and Russia on board to support the UN sanctions despite North Korea's threat that the sanctions equaled war. No one was willing to provide a meaningful third-party deterrence to protect Pyongyang's nuclear ambitions.

Playing diplomacy with Russia and China was not the only difficult task U.S. diplomats had to manage – dealing with U.S. allies often appeared more troubling. European nonproliferation hawks were reliable supporters and could provide useful help, but apart from London's and Paris's votes in the Security Council their support was not vital. But fighting a war against North Korea would be impossible without Japan and the ROK. As Gallucci recalls, "[f]or the Clinton administration, the politics in Washington and with Seoul were often as challenging as dealing with the North."[92] The ROK was particularly difficult to deal with. Many South Koreans were sympathetic to North Korea for a variety of reasons ranging from having relatives there, to having leftist sympathies, to disliking the continued U.S. presence in their country. Thus a deep division on policy toward North Korea between ROK's hawks and doves, both regularly playing the issue in domestic politics, complicated things further. American officials regularly complained about Seoul's unpredictability – one moment Seoul was complaining the Americans were too weak and the next that they were too aggressive.[93] When both Tokyo and Seoul were consulted about a possible strike in May 1993, they were unenthusiastic.[94] Yet as the concerns over the DPRK's behavior grew and U.S. diplomatic efforts came to fruition both governments relented by June 1994, turning to supporting the U.S. plan for UN sanctions despite the threat that the move would lead to war.

Securing support for UN action from the broadest possible coalition was important for legitimizing the U.S policy toward the DPRK to an international audience. But making the action against North Korea appear legitimate was not a particularly difficult task. The DPRK's self-imposed isolation, the increasing legitimacy both of international interventions and non-proliferation efforts, and the weakening of the non-intervention norm all played a role. Mostly only states that were themselves considered on the verge of being international outlaws were ready to support Pyongyang. When the vote took place on the DPRK's non-cooperation with the IAEA in the UN General Assembly in November 1993, only 10 states abstained and only one – North Korea itself – refused to support the resolution calling on Pyongyang to cooperate.[95] While this does not necessarily mean consent with U.S. military action, clearly the DPRK's behavior was illegitimate enough to the international audience to suppress any protection the non-intervention norm could have provided.

Turning to another international norm possibly constraining U.S. policy, it appears that by 1994 the nuclear taboo may have been strong enough to influence U.S. behavior, but arguably had little practical impact on the decision to strike Yongbyon. The available evidence suggests that the possibility of striking first with U.S. nuclear weapons was not even discussed by the relevant decision makers, despite possible advantages such weapons might have had. Yet it is true that, if it had been discussed, it would turn out that the advantage of a nuclear strike would be dubious. As Jasper Becker observes, with the satellite imagery of the targets at Yongbyon, their destruction with cruise missiles and F-117 stealth fighters dropping laser-guided bombs seemed simple.[96] Conventional explosives would easily do the job; there was no need to have nuclear weapons involved. More concerning to decision makers was the danger that destroying the DPRK's reactor might release radiation, possibly on Japan.[97] Yet in this case, the reasons do not appear to be normative, but rather utilitarian. How the aforementioned would influence the decision on intervention will remain unclear. Jimmy Carter's intervention stopped the planning early enough. Still, it is reasonable to say that a strong nuclear taboo existed with respect to the first strike with nuclear weapons, and striking a hot reactor appeared to be dangerous. Nonetheless, the danger might have been dealt with or simply accepted, or at least this is what Secretary Perry suggested when testifying to Congress in 1995 on the subject of a pre-emptive strike against Yongbyon, while the nuclear first strike was simply unnecessary.[98]

What certainly was not limiting Clinton's ability to destroy Yongbyon was the domestic audience. In fact, the relevant audience was largely more hawkish than the administration. Six out of 10 Americans were afraid that U.S. vital interests were at stake in North Korea.[99] In general, the Clinton administration was perceived as being too soft on the North Koreans. Domestic politics occasionally even made the administration take a tougher stance on the DPRK. Conservative columnists, seconded by hawkish members of Congress, were particularly eager to call for the strike.[100] When the final deal defusing the crisis – the Agreed Framework – was concluded between the United States and the DPRK, domestic objections were widespread. They even increased when the Republicans conquered the Congress in the 1994 mid-term elections.[101] Those who opposed to the Agreed Framework complained that the DPRK would get benefits and cheat in its nuclear program, calling for resolute action.[102] Certainly had the administration decided to strike, it would have found enough supporters rallying around its flag.

It would not even have found decisive opposition within its own ranks. On the one hand, apparently there was not unity in the Clinton administration. In particular, the rift went between Pentagon and the State Department. While the former's priority was to crumble further proliferation regardless of what happened in the past, the latter put emphasis on preserving the nonproliferation regime, thereby requesting full disclosure of North Korea's nuclear past.[103] On the other hand, if the event got to the last-resort moment (and it was agreed that the strike should be postponed for then), neither sides' goals would be significantly hampered further.

Considering the regime type factor, the U.S. in 1994 is predictably easy to describe. While every democracy has unavoidable imperfections, the U.S. certainly represents the best-performing ones, scoring a full 10 points in the Polity IV database. Turning to the role of the military, it must be noted that Clinton's relations with the armed forces were uneasy, particularly in the beginning of his presidency. His halfhearted decision on the rights of homosexuals in the armed forces, known as the "don't ask don't tell" rule, did not satisfy gay rights advocates but did upset the commanders.[104] Furthermore, the mismanagement in Somalia in 1993 that left 18 American servicemen dead and the United States humiliated when their bodies were dragged in the streets of Mogadishu made the administration "strongly inclined to support all requests from field commanders."[105] Yet despite the military's strong voice over professional matters, it is unreasonable to judge Clinton's decision making as over-influenced by his military commanders. At the end of the day, the military appeared ready to bomb Yongbyon if it was asked to and, if requested, reinforcements were transferred to Korea. The vision of their civilian leaders did not differ.

As with the previous case studies, the critical part comes particularly in accessing the effectiveness of the four possible deterrence threats. With the intelligence estimate that there was a better-than-even chance Pyongyang had enough plutonium for one or two nuclear weapons, possibly even the bomb itself, it comes as little surprise that any clues of sensitivity to the nuclear denial threat are absent in what we know about U.S. decision making. Apparently one or two can hardly deny much. Yet uncertainty should have made the threat of nuclear retaliation work, or at least that is what a proliferation optimist would expect.

However, it appears that the sensitivity to the DPRK's nuclear retaliation was relatively low. A *New York Times* editorial brought a rare warning, calling that "if intelligence assessment is correct ... Pyongyang's response could be disastrous."[106] But this view was not widely shared by the decision makers. The administration may have had this threat in mind, but it is uncertain that it considered it prohibitive. At the end of the day, it would have been difficult for the DPRK to strike targets valuable to Washington. Few precautions were taken to mitigate the threat of the DPRK's nuclear retaliation, even though it was possible to take some. Arguably the best explanation is that there was a much more robust threat in the DPRK's arsenal occupying the minds of U.S. planners.

As has been already noted, the DPRK was threatening an escalatory response – conventional retaliation – to U.S. action. This could mean a new fully-fledged war on the peninsula where Seoul was precariously vulnerable to North Korea's voluminous artillery. Such a war had frightening prospects. A coalition victory appeared certain, but with high costs. Worst-case estimates reached a million causalities, including 80,000–100,000 U.S. military causalities, and a U.S.$100 billion price tag.[107] U.S. commander in Korea General Luck informed Daniel Poneman from the National Security Council that a recent war game showed "the United States and South Korea would win, but with 300,000–750,000 casualties among military personnel. That did not include civilian casualties or measure the enormous economic damage to the South."[108] The sensitivity to the DPRK's conventional might is well captured by U.S. Ambassador in Korea Laney's account that he and General Luck agreed the DPRK's nuclear program was not worth risking a conventional war. "Why are we going to risk killing a million people? A bomb or two can't even do that," Laney and Luck asked themselves.[109] The administration agreed that precautions should go in this direction. The critical precondition to strike, as formulated by the Principals Committee in its last session before Carter's deal with Kim Il-Sung, was to bring in the reinforcements to tackle the DPRK's attack. Clearly the U.S. sensitivity focused on conventional retaliation.

Conventional denial was far less concerning. North Korea's nuclear complex was well protected by 300 antiaircraft guns and six SAM missile sites. Furthermore, a hilly terrain would complicate the attack.[110] Yet none of that would be prohibitive, the U.S. air force was far too advanced in its development. As Becker observed, F-117 stealth ground-attack aircraft and cruise missiles could easily accomplish the destruction of Yongbyon.[111] The only difficulty would be in destroying targets of unknown location.

The results are summarized in Table 5.1 in an illustrative fashion. Not dissimilar to the other deterrence successes that have been presented so far, this chapter most importantly reveals how the threat of conventional retaliation served as a major stabilizer in a dyad that was apparently destabilized by the emergence of a new nuclear arsenal. Key to deterrence success was the DPRK's conventional preponderance in the theater, accompanied by the image of the country's strong resolve. Yet, whether deterrence could have held when other options would have been exhausted in 1994 will never be fully answered. While the DPRK's conventional threat was apparently quite robust, the injection of U.S. reinforcements could have undermined it with technologically sophisticated anti-artillery capabilities, raising both the dynamics and the limits of conventional deterrence.

Table: 5.1 U.S.–DPRK, 1992–1994

Concept	Description	Values
The deterrer's nuclear arsenal	Key physical attributes of respective nuclear complexes; weapon types; numbers; delivery vehicles; command, control, and communication systems	0–2 (better-than-even chance that North Korea had one or two plutonium weapons; no early warning system; primitive command and control)
Nuclear asymmetry	Situation when a substantially smaller (quantitatively) and less sophisticated nuclear force (qualitatively) faces a qualitatively and quantitatively larger nuclear force	Strong asymmetry (quantitative superiority by a factor of at least 1,000 and a comparable qualitative edge)
Second-strike criterion	Nuclear posture that has the ability to survive enemy attack, make and communicate decision to retaliate, overcome enemy's active defense, and destroy a valuable target despite its passive defense	None (weapons would lack delivery vehicles)
General conventional preponderance	Situation when one side's armed forces are in general substantially stronger in terms of numbers, technology, training, and employment strategies	Challenger (unequivocally stronger navy, air force, and land forces; edge in terms of quality and quantity)
Theater conventional preponderance	Situation when one side's armed forces are substantially stronger in terms of numbers, technology, training, and employment strategies on a theater where targets valuable to the challenger can be found	Deterrer (short-term advantage against Seoul; if not long enough for a breakthrough than for destruction of the city by artillery fire)
The challenger's technological advantage	Significant advantage of the challenger's major weapons systems that would be employed in case of conflict in terms of state-of-the-art sophistication over the deterrer's weapons systems	Strong (significant edge in sophistication of most weapon platforms; some technologies completely unavailable to the deterrer)

Availability of information	The challenger's knowledge about targets' location and defensive systems, which is established from sources that the challenger deems credible	Limited (impressive overhead imagery, but poor HUMINT and imperfect cooperation with the ROK's allied intelligence)
Centrality of conflict	Absolute importance of the conflict dyad between the challenger and the deterrer from the challenger's subjective perspective and its relative importance to other existing conflict dyads where the challenger is a party	Yes (limited prior to the 1993/1994 period; then full attention paid by top leaders)
Perceived resolve of the deterrer	The challenger's perception of the deterrer's commitment to fight should the deterrence fail	Strong (decades of hostile behavior with little sensitivity to punishment)
Institutionalization of mutual relations	Degree of shared expectations about the requirements of stable deterrence and the existence of proven formal or informal communication channels	Limited (almost no contact prior to 1990; then several meetings but with little time to establish common understanding)
History of hostility	Track record of conflict in the dyad that shapes the understanding of other side's intentions	Strong (dating back to the Korean War; effectively prohibiting U.S. ability to talk with Kim Il-Sung directly)
Last resort	Situation when the challenger sees only the options to strike, or to live up to the development he tries to prevent	Limited (sanctions believed to be the next step before bombing)
Perception of the deterrer's rationality	The challenger's perception of the deterrer's rationality, particularly whether the challenger believes it possible to live with the nuclear-armed deterrer in the long term	No (the basic assumption was that the North Koreans were liars, dug tunnels and could not be trusted)

continued

Table 5.1 Continued

Concept	Description	Values
Third-party deterrence	Threat of military involvement into the original conflict by a third party, most likely the allies of the original deterrer, that decisively influenced the challenger's decision	Limited (DPRK's allies eased their commitments and were moving in favor of sanctions)
Alliance politics	Sensitivity of the challenger to possible impact of his action on the relations with his allies	Strong (large sensitivity to ROK's and Japan's position, both slowly taken on board)
International action legitimacy	The challenger's sensitivity to the international normative expectation of non-intervention	Limited (successful coalition building effort provided legitimacy; normative shield undermined by DPRK's outlaw reputation)
Nuclear taboo	Normative prohibition on the use of nuclear weapons	Strong (taboo already well-established in the U.S. but not practically influencing decisions, with conventional weapons sufficient for the mission)
Domestic action legitimacy	Level of support for military solution among the challenger's population	Strong (domestic audience more hawkish than administration)
Opposition among influential decision makers	Negative view on military solution by influential part of the challenger's government	Limited (consensus on last resort)
Regime type	The challenger's position on the democracy-nondemocracy axis and on the militarized-nonmilitarized axis	Democracy, strong civilian control of armed forces

Nuclear retaliation	Threat that the deterrer will use its nuclear weapons against targets valuable to the challenger, except of targets that are directly related to pursuit of the challenger's objectives	Limited (no concerns about the DPRK's nuclear retaliation raised in the government; no preparatory measures taken to mitigate such threat)
Nuclear denial	Threat that the deterrer will use its nuclear weapons against targets that are directly related to pursuit of the challenger's objectives in order to prevent him from attaining the objectives, or in order to make it unacceptably costly	Limited (no concerns about the DPRK's nuclear denial raised; no preparatory measures taken to mitigate such threat)
Conventional denial	Threat that the deterrer will use its conventional weapons against targets that are directly related to pursuit of the challenger's objectives in order to prevent him from attaining the objectives, or in order to make it unacceptably costly	Limited (some concerns about anti-aircraft protection of targets and unknown location of plutonium raised, but not considered prohibitive)
Conventional retaliation	Threat that the deterrer will use its conventional weapons against targets valuable to the challenger, except of targets that are directly related to pursuit of the challenger's objectives	Strong (strong concerns about the DPRK's attack on the South raised; conventional build-up prepared)

Notes

1 Walter C. Clements, "North Korea's Quest for Nuclear Weapons: New Historical Evidence," *Journal of East Asian Studies*, 10/1 (2010) 127–154.
2 Don Oberdorfer, *The Two Koreas: A Contemporary History* (New York: Basic Books, 2001) 304.
3 Joel S. Wit, Daniel B. Poneman, and Robert L. Gallucci, *Going Critical: The First North Korean Nuclear Crisis* (Washington D.C.: Brookings Institution Press, 2004) 1.
4 Leon V. Sigal, *Disarming Strangers: Nuclear Diplomacy with North Korea* (Princeton: Princeton University Press, 1999) 25.
5 Jasper Becker, *Rogue Regime: Kim Jong Il and the Looming Threat of North Korea* (Cary: Oxford University Press, 2006) 166.
6 Sigal, *Disarming Strangers*, 27–30.
7 John F. Farrel, "Team Spirit: A Case Study on the Value of Military Excercises as a Show of Force in the Aftermatn of Combat Operations," *Air & Space Power Journal*, 23/3 (Fall 2009) 99.
8 Oberdorfer, *The Two Koreas*, 321–323.
9 Sigal, *Disarming Strangers*, 32, 39.
10 Oberdorfer, *The Two Koreas*, 324.
11 Becker, *Rogue Regime*, 184.
12 Sigal, *Disarming Strangers*, 43.
13 Sigal, *Disarming Strangers*, 46.
14 Wit, Poneman, and Gallucci, *Going Critical*, 19–21.
15 Wit, Poneman, and Gallucci, *Going Critical*, 24; Oberdorfer, *The Two Koreas*, 336.
16 Wit, Poneman, and Gallucci, *Going Critical*, 26–29.
17 Sigal, *Disarming Strangers*, 57.
18 Sigal, *Disarming Strangers*, 57.
19 Wit, Poneman, and Gallucci, *Going Critical*, 32.
20 Wit, Poneman, and Gallucci, *Going Critical*, 43–44.
21 Sigal, *Disarming Strangers*, 62.
22 Wit, Poneman, and Gallucci, *Going Critical*, 45–47.
23 Wit, Poneman, and Gallucci, *Going Critical*, 48–50.
24 Wit, Poneman, and Gallucci, *Going Critical*, 51–63; Sigal, *Disarming Strangers*, 64; Oberdorfer, *The Two Koreas*, 338–343.
25 Wit, Poneman, and Gallucci, *Going Critical*, 74; Oberdorfer, *The Two Koreas*, 347.
26 Sigal, *Disarming Strangers*, 72–73.
27 Wit, Poneman, and Gallucci, *Going Critical*, 84–88.
28 Wit, Poneman, and Gallucci, *Going Critical*, 107.
29 Wit, Poneman, and Gallucci, *Going Critical*, 115–117; Sigal, *Disarming Strangers*, 98–99.
30 Oberdorfer, *The Two Koreas*, 356–357.
31 Sigal, *Disarming Strangers*, 104.
32 Wit, Poneman, and Gallucci, *Going Critical*, 143–144.
33 Sigal, *Disarming Strangers*, 107.
34 Wit, Poneman, and Gallucci, *Going Critical*, 149.
35 Sigal, *Disarming Strangers*, 107.
36 Wit, Poneman, and Gallucci, *Going Critical*, 154.
37 Wit, Poneman, and Gallucci, *Going Critical*, 154.

38 Sigal, *Disarming Strangers*, 110.

39 Sigal, *Disarming Strangers*, 117.

40 Wit, Poneman, and Gallucci, *Going Critical*, 182–191.

41 Samuel S. Kim, "North Korea in 1994: Brinkmanship, Breakdown, and Break-through," *Asian Survey*, 35/1 (January 1995) 21.

42 Wit, Poneman, and Gallucci, *Going Critical*, 210.

43 Wit, Poneman, and Gallucci, *Going Critical*, 166.

44 Wit, Poneman, and Gallucci, *Going Critical*, 179.

45 Wit, Poneman, and Gallucci, *Going Critical*, 210.

46 Oberdorfer, *The Two Koreas*, 371.

47 Sigal, *Disarming Strangers*, 155.

48 Wit, Poneman, and Gallucci, *Going Critical*, 226.

49 Wit, Poneman, and Gallucci, *Going Critical*, 226–231.

50 Wit, Poneman, and Gallucci, *Going Critical*, 244.

51 Jonathan D. Pollack, "The United States, North Korea, and the End of the Agreed Framework," *Naval War College Review*, 56/3 (Summer 2003) 12.

52 Sigal, *Disarming Strangers*, 41; Wit, Poneman, and Gallucci, *Going Critical*, 54.

53 Wit, Poneman, and Gallucci, *Going Critical*, 38.

54 Becker, *Rogue Regime*, 166.

55 Wit, Poneman, and Gallucci, *Going Critical*, 128.

56 Robert S. Norris and William M. Arkin, "U.S. Strategic Nuclear Forces, End of 1992," *The Bulletin of Atomic Scientists*, 49/1 (January/February 1993) 57.

57 Sigal, *Disarming Strangers*, 170.

58 Sigal, *Disarming Strangers*, 22.

59 Lawrence Freedman and Efraim Karsh, "How Kuwait Was Won: Strategy in the Gulf War," *International Security*, 16/2 (Fall 1991) fn. 65.

60 John M. Collins, "Korean Crisis, 1994: Military Geography, Military Balance, Military Options," *CRS Report for Congress*, No. 94–311S, April 11, 1994.

61 Sigal, *Disarming Strangers*, 21.

62 Victor D. Cha, "North Korea's Weapons of Mass Destruction: Badges, Shields, or Swords?," *Political Science Quarterly*, 117/1 (Summer 2002) 226.

63 Sigal, *Disarming Strangers*, 74.

64 Wit, Poneman, and Gallucci, *Going Critical*, 107.

65 Wit, Poneman, and Gallucci, *Going Critical*, 165.

66 Clements, "North Korea's Quest for Nuclear Weapons," 145.

67 Wit, Poneman, and Gallucci, *Going Critical*, 104.

68 See Stephen Biddle, "Victory Misunderstood: What the Gulf War Tells Us about Future of Conflict," *International Security*, 21/2 (Autumn 1996) 139–179.

69 Sigal, *Disarming Strangers*, 234.

70 Clements, "North Korea's Quest for Nuclear Weapons," 103.

71 Oberdorfer, *The Two Koreas*, 371.

72 Wit, Poneman, and Gallucci, *Going Critical*, 17.

73 Wit, Poneman, and Gallucci, *Going Critical*, 48.

74 Sigal, *Disarming Strangers*, 212.

75 Wit, Poneman, and Gallucci, *Going Critical*, 188.

76 Robert Jervis, "The Impact of the Korean War on the Cold War," *The Journal of Conflict Resolution*, 24/4 (December, 1980) 563–592.

77 Pollack, "The United States, North Korea, and the End of the Agreed Framework," 15.

78 Roland Bleiker, "A Rogue Is a Rogue Is a Rogue: U.S. Foreign Policy and the Korean Nuclear Crisis," *International Affairs*, 79/4 (July, 2003) 736.

79 Wit, Poneman, and Gallucci, *Going Critical*, 132.
80 Peter Howard, "Why Not Invade North Korea? Threats, Language Games, and U.S. Foreign Policy," *International Studies Quarterly*, 48/4 (2004) 820.
81 Sigal, *Disarming Strangers*, 230.
82 Sigal, *Disarming Strangers*, 10.
83 Wit, Poneman, and Gallucci, *Going Critical*, 52
84 Morgan, Patrick M., "Deterrence and System Management: The Case of North Korea," *Conflict Management and Peace Science*, 23/2 (2006) 128.
85 Morgan, "Deterrence and System Management," 129.
86 Sigal, *Disarming Strangers*, 54.
87 Wit, Poneman, and Gallucci, *Going Critical*, xiv.
88 Cha, "North Korea's Weapons of Mass Destruction," 218.
89 Wit, Poneman, and Gallucci, *Going Critical*, 156.
90 Sigal, *Disarming Strangers*, 118.
91 Sigal, *Disarming Strangers*, 58.
92 Wit, Poneman, and Gallucci, *Going Critical*, xv.
93 Wit, Poneman, and Gallucci, *Going Critical*, 114
94 Sigal, *Disarming Strangers*, 118.
95 Sigal, *Disarming Strangers*, 73.
96 Becker, "Rogue Regime," 165.
97 Sigal, *Disarming Strangers*, 335; Wit, Poneman, and Galucci, *Going Critical*, 103.
98 Wit, Poneman, and Galucci, *Going Critical*, 103.
99 Wit, Poneman, and Galucci, *Going Critical*, 241.
100 Sigal, *Disarming Strangers*, 71.
101 Pollack, "The United States, North Korea, and the End of the Agreed Framework," 19.
102 Robert M. Hathway and Jordan Tama, "The U.S. Congress and North Korea during the Clinton Years: Talk Tough, Carry a Small Stick," *Asian Survey*, 44/5 (September/October 2004) 711–733.
103 Wit, Poneman, and Galucci, *Going Critical*, 140.
104 Richard H. Kohn, "Coming Soon: A Crisis in Civil–Military Relations," *World Affairs*, 170/3 (Winter 2008) 69.
105 Wit, Poneman, and Gallucci, *Going Critical*, 122.
106 Sigal, *Disarming Strangers*, 94
107 Sigal, *Disarming Strangers*, 335.
108 Wit, Poneman, and Gallucci, *Going Critical*, 102.
109 Sigal, *Disarming Strangers*, 122.
110 Wit, Poneman, and Gallucci, *Going Critical*, 103.
111 Becker, "Rogue Regime," 165.

6 The United States and the Soviet Union, 1962

The story about "the missiles of October" is so notoriously well known that it almost does not need retelling. Leading Cold War historian John Lewis Gaddis is not the only one to observe that "no episode in the history of international relations has received such microscopic scrutiny from so many historians" and that "theorists have generalized exuberantly from this single specific event."[1] Possibly there is not much new to say about the Cuban Missile Crisis. It may be bad for an ambitious historian trying to tell something new. But it certainly is good for a student of deterrence. With the abundance of scholarship on the Cuban Missile Crisis and on the role nuclear deterrence played in the course of events, this is probably the closest to Archimedes' "place to stand" students of deterrence can currently get. And it certainly is a useful case for comparison.

The plot

On October 16, 1962, President Kennedy was presented with new and extremely disturbing overhead imagery. Taken two days before by a U-2 overflight, the photos showed Soviet MRBM and IRBM missile sites under construction in Cuba. The President was surprised and outraged. He had publicly warned the Soviets in September not to place offensive weapons in Cuba.[2] The Soviets including Khrushchev himself had assured him that they would not do it. Under such circumstances it was impossible to accept the Soviet deployment. Kennedy quickly summoned his closest national security advisors to debate U.S. reaction. With the President's preference of an informal working style, the membership of the group fluctuated and did not closely follow the formal responsibilities and positions of the persons involved. The key members included Secretary of Defense Robert McNamara with his deputy Roswell Gilpatric and Assistant Secretary Paul Nitze, chairman of the JCS Maxwell Taylor, Secretary of State Dean Rusk with Under Secretary George Ball and Assistant Secretary for Inter-American Affairs Edwin Martin, the President's national security advisor McGeorge

Bundy, the President's speechwriter Theodore Sorensen, former ambassador to Moscow Llewellyn Thompson, Director of Central Intelligence John McCone, Secretary of Treasury Douglas Dillon, former Secretary of State Dean Acheson, and Attorney General Robert Kennedy. The group would be known as the ExCom (Executive Committee of the National Security Council) and, starting from October 16, it debated policy options including that of destroying Soviet missiles with an air strike.[3]

With the benefit of hindsight, Kennedy should not have been so surprised by the Soviet deployment. By Fall 1962, many in the United States were deeply suspicious of Soviet intentions. In July, the intelligence detected Raul Castro's arrival in Moscow. By August it was known that substantial deliveries of Soviet equipment and men were flowing from the Soviet Union to Cuba. At the end of August, the U-2s spotted the positions of SA-2 missiles. It was likely that the SA-2s were deployed to protect something valuable. DCI McCone rightly estimated that the valuable things were nuclear missiles, but he was the only one.[4] There was an equally good reason for skepticism. The administration had clearly signaled to the Soviet Union that missile deployment would be unacceptable, or at least the Americans believed so. In September, the Soviets were further warned not to place offensive weapons in Cuba. They understood what the Americans meant but the missiles' deployment was already underway. Furthermore, Khrushchev's motives were defensive and accordingly he did not see the missiles for Cuba as offensive weapons. He was genuinely afraid that the Americans were preparing to get rid of Castro's regime with a full-scale invasion. The United States would not understand his fears. They would see the Soviet move as a challenge to their resolve.[5] When the crisis was over, the historians assumed that it arose from the Soviet need to offset their inferiority in strategic weapons by placing the IRBMs and MRBMs they had in abundance in Cuba. Only after the end of the Cold War was it discovered what Khrushchev's intentions were.[6]

Soon after the ExCom first met on October 16, it was clear that no one considered it acceptable to simply leave the Soviet missiles where they were. The President was oversensitive about appearing irresolute to foreign and domestic audiences after the Bay of Pigs fiasco, and Cuba was his "Achilles heel."[7] The airstrike quickly developed as the preferred option of the majority of ExCom members, but practical issues of concern emerged. McNamara argued that an airstrike could take place only before the missiles became operational. Thereafter some might survive and be launched against the United States.[8] General Taylor advised striking the missiles without warning to obtain strategic surprise, but he had to admit that even such a strike would not destroy 100 percent of the missiles. From his military perspective Taylor also asked to take out the SAM batteries together with the missile sites to protect U.S. aircraft.[9]

After a few hours the ExCom meeting had to be postponed to the evening. In the meantime, the Joint Chiefs of Staff met to discuss the issue as well. Their recommendation was to take out not only the missiles but a range of other targets as well in a large surprise strike.[10] The ExCom then gathered again. No real decision was taken in its fairly unstructured discussion, but it was largely agreed that the options should be prepared in more detail.[11]

An important question emerged about the nature of the contemplated airstrike. It had to be decided what to target. It was obvious to attack the missiles themselves, but should the U.S. also strike Cuban airfields, SAM batteries and coastal missile sites, or other significant targets? Striking without warning would be of military benefit, particularly should the strike be kept surgical, but many opposed this due to political costs and moral objections. Even hawks like McCone and Dillon, who were in favor of an airstrike, objected to a Pearl-Harbor-style attack. The less hawkish consented. George Ball was concerned about Soviet reaction and the moral downside of a surprise attack. Importantly, Robert Kennedy backed Ball's moral objections to a surprise attack. Yet warning the Soviets in advance would let them prepare their defenses and would practically make it impossible to keep the strike limited to the missiles. The Joint Chiefs staunchly supported the surprise option, urging McNamara on October 17 not to limit the strike and not to warn the Soviet defenses in Cuba in advance. But by October 18 support for the airstrike faded.[12]

Now a naval blockade, first suggested by McNamara on October 16, became the preferable option. Initially the blockade was viewed as more extreme than the airstrike, since traditionally it was considered an act of war under international law. But the Department of Justice crafted the concept of a "visit-and-search" blockade that would be limited to offensive weapons heading to Cuba. It would still require searching Soviet ships on the high seas, but it appeared less risky than killing Soviet soldiers with an airstrike. Furthermore it was believed that allies could be relied on to support the limited blockade, while they would certainly oppose the airstrike. Llewellyn Thompson took the lead in advocating the limited blockade and this option gathered support from key figures including Secretaries McNamara and Rusk, and the President's brother, Robert. The military, backed by McCone, Dillon, Acheson, and now Bundy, still supported striking as soon as possible, arguing that the blockade would only allow the Soviet missiles to become operational. Yet the group was weakened when Acheson decided to leave the fight as he felt support for the airstrike could not be won. All remaining supporters of the airstrike were in favor of a more extensive attack not limited to the missile sites, but this was politically more difficult to sell.[13] On October 20, the President made the decision to authorize the blockade.[14] Yet the decision had a downside. It could not take the missiles out of Cuba. Certainly it was hoped that the Soviets would withdraw them in response to the blockade, but no decision was made about what to do if they did not.

On October 22, President Kennedy announced in a televised address a "naval quarantine" around Cuba. No offensive weapons and associated material would be allowed to continue to Cuban ports. The U.S. Navy deployed 183 ships to enforce the blockade. Simultaneously, the armed forces started to prepare for a possibility that the blockade would not work. Five Army and one Marine divisions, some 140,000 troops, were gathered in Florida to prepare for a possible invasion.[15] To make sure the Soviets understood the importance of the moment, Robert Kennedy met with Soviet Ambassador Dobrynin. But the initial Soviet reaction was worrying; the younger Kennedy learned that Soviet ships were to carry on.[16] Fortunately, Khrushchev was not interested in provoking a military action. The Soviet ships were ordered to turn back or stop. Later Khrushchev moderated his decision and decided to test the Americans. A Soviet tanker, Bucharest, challenged the blockade. Since such a ship was unlikely to carry offensive weapons, the Americans let Bucharest go through, only visually inspecting it from afar but not boarding.[17]

Things appeared to be going well but soon turned worse. On October 26, a private letter arrived from Khrushchev, offering that the Soviets would remove the missiles from Cuba in exchange for the Americans' pledge not to invade the island. "The initial reaction was that this was too good to be true," observed Lawrence Freedman.[18] No U.S. invasion was planned anyway so the crisis would be defused at a low cost. But on the morning of October 27, things became complicated. Moscow broadcasted another massage that now included a tit-for-tat demand for the U.S.: the Soviet missiles in Cuba for U.S. missiles in Turkey. This was not the only bad news coming to Washington. Another Soviet ship was now approaching the quarantine line, the construction of the missile sites in Cuba was progressing rapidly, and the missiles were becoming operational. The already heated situation was then pushed further when the report came that a U-2 was shot down over Cuba and that another one was pursued by Soviet fighters over Siberia.[19] The hawks called for a reciprocal attack against the Soviet SAM battery responsible for downing the American aircraft, arguing that restraint would undermine the American position and prevent further intelligence gathering.

The ExCom discussed the situation during its critical meeting on October 27. The precarious question of how to respond to Khrushchev's two letters had to be addressed. As it is well known, Kennedy decided to ignore the second message and respond to the first one. The fate of the Turkish missiles was also thoroughly discussed by the ExCom. The President was in favor of trading the militarily useless missiles in Turkey for defusing the crisis, but it was clear that the explicit trade would enrage allies.[20] No mention of the Turkish missiles was included in the official response to Khrushchev, which was publicly aired. However, Robert Kennedy was sent to meet in private with the Soviet ambassador. The Attorney General presented Dobrynin with what can be seen as a de facto ultimatum. The Soviets would have to agree to withdraw their missiles, otherwise pressures in the U.S. government would make military action to remove them inevitable. However, there was a sweetener for the Soviets. The American missiles would be quietly removed from Turkey within four to five months, but the Soviets could not even hint such a trade in public to allow the withdrawal appear unrelated to the Cuban events.[21]

Under such circumstances Khrushchev accepted withdrawing the missiles. He was worried about the situation getting out of hand. The Soviet troops in Cuba were not allowed to fire on American reconnaissance planes yet they shot down the U-2.[22] In Moscow, the Soviet leaders discussed the possible responses to a U.S. attack on Cuba including action against West Berlin or the U.S. missiles in Turkey. This was exactly the kind of Soviet retaliation Washington was afraid of. But the Soviet conclusion was that such options would provoke escalation. They were even seriously afraid that the U.S. might decide to pre-empt the Soviet Union.[23] With the concession on Turkish missiles, Khrushchev hastened to accept the deal. The Americans could claim a great victory. Khrushchev would be seen as a loser, but he had few reasons to feel defeated. Protecting Castro from an American invasion was his primary rationale for sending the missiles. The move was not particularly well thought out, but defensive in nature. In return for a withdrawal he got the U.S. pledge not to invade Cuba and to withdraw missiles from Turkey in a short space of time.[24] Furthermore, war was avoided.

Unfolding the complexity

Certainly more interesting than retelling the story of the Cuban Missile Crisis is placing it into a comparative framework. It should not be difficult to see that it is a useful comparison. In 1962, the Soviet Union was a global superpower; its impressive tank armies regularly haunted Western decision makers, its nuclear arsenal was rapidly growing in numbers. While it was yet far from being on a par with the United States, the quality of asymmetry was markedly different from other cases. This makes the case reasonable to use as a control when the effectiveness of both nuclear and conventional deterrence and various aspects influencing them are the researcher's central interest.

Just a few years before the Cuban crisis it appeared that the Soviet Union's nuclear arsenal was rapidly catching up with or even surpassing the United States. After the launch of Sputnik in 1957, Khrushchev publicly bragged about his new strategic capabilities.[25] Only recently has U.S. intelligence confirmed that the missile gap was a big bluff. Khrushchev's allegedly fearsome ICBMs were few in numbers and poor in quality. In reality, the Soviets had only six huge, yet virtually useless "zero generation" SS-6 ICBMs and 36 slightly more sophisticated, but still highly vulnerable first generation SS-7 ICBMs.[26] The U.S. intelligence estimate at the time – 75 Soviet ICBMs – was still overstated.[27] The other legs of the Soviet strategic triad were also far from advanced. The Soviet air force deployed 100 TU-95 Bear turboprop and 60 3M Bison jet strategic bombers with 270 operational nuclear warheads.[28] Starting in 1961, the Soviet navy also deployed its first ballistic missile submarines with SS-N-4 missiles. By the time of the Cuban Missile Crisis, 66 SS-N-4 were deployed in the Soviet navy but the missiles had a limited range, unimpressive accuracy, and could be only launched when the submarine surfaced.[29] Notwithstanding its relative unsophistication, the Soviet strategic arsenal was still far from small and it was coupled with an impressive number of sub-strategic nuclear weapons. More than 500 SS-4 MRBMs were deployed in Europe, together with 28 SS-5 IRBMs and an unknown number of nuclear-capable tactical aircraft. In Cuba, the Soviets had 42 SS-4 MRBMs and 32 SS-5 IRBMs; but only six to eight became operational during the crisis. The Soviet troops there also had 12 FROG short-range ballistic missiles and 80 cruise missiles for the coastal defense.[30]

The Soviets were well aware that their arsenal was vulnerable and less sophisticated then the American one.[31] The U.S. had 121 Atlas, 53 Titan-liquid fueled ICBMs, and eight modern solid-fueled Minuteman ICBMs. The Navy had already deployed its first highly survivable nuclear-powered ballistic missile submarines carrying 112 modern Polaris SLBMs. Most importantly, the U.S. Air Force's Strategic Air Command had an impressive number of strategic bombers. Some 880 B-47s, 639 B-52s, and 76 B-58s served in the air force, carrying 2,952 nuclear warheads. In the European theater the Soviet forces faced 45 Jupiter IRBMs in Turkey and Italy and 60 Thor IRBMs in Britain, all capable of hitting targets in the Soviet Union. Furthermore, NATO deployed a large number of various tactical nuclear weapons, planning to use them to offset its conventional inferiority.[32] The U.S. arsenal enjoyed an appreciable edge in quality and quantity though, yet the asymmetry was far from any of the previous cases.

The U.S. superiority combined with the vulnerability of the early Soviet nuclear weapons platforms put the survivability of the Soviet strategic arsenal and its ability to retaliate in question. No early warning system worthy of that name was available to the Soviets. By 1962, U.S. satellites were in service over the Soviet Union. Washington – largely successfully – devoted a substantial effort into locating Soviet ICBMs, not least to confirm or dismiss the existence of the missile gap. Once the location was pinpointed by U.S. satellites and U-2s, Soviet ICBMs – liquid-fueled, slow to launch, not deployed in hardened silos – became vulnerable to the U.S. first strike. The same applied to the Soviet bomber force, though at least there could have been some confidence in Moscow with numbers. However, the surviving bombers would have to be launched against a formidable U.S. air defense, making a successful second strike against U.S. soil uncertain.[33] Also the Soviet submarines did not fill them with confidence. They were considerably louder than their Western counterparts and accordingly more vulnerable to anti-submarine warfare. The missiles they carried had a short range and could not be fired submerged. The few operational MRBMs in Cuba that could reach the United States could be destroyed by U.S. air strikes. It is impossible to state how many deliverable strategic forces would be left to the USSR after the U.S. first strike. Cimbala estimates that number at 41 warheads, which would then have to be launched against the U.S. with a high probability of at least some of them being lost on that mission.[34] The prospects of the U.S. first strike were good, but did not lend themselves to full confidence; as General Taylor urging for the air strike on Cuba had to admit, the U.S. would be "pushing a 100 percent [success rate] just as far, as closely as we can with our, with our airstrike."[35] Taylor could not say with confidence that all the Soviet missiles in Cuba would be destroyed, and destroying all of them together with Soviet ICBMs, bombers and nuclear-armed submarines would be even more difficult. Even if it could be achieved, there was no chance of knocking out all Soviet sub-strategic weapons.

Ever since the end of the Second World War, the conventional preponderance of Soviet tank armies – at least in the Cold War's central theater in Europe – was common knowledge. Yet, in fact, global superiority belonged to the United States. Lacking real global power projection capabilities well into the 1970s, the Soviets could only fight in Europe.[36] There the Warsaw pact had 175 divisions against only 25 of NATO. At first glance, this appeared more than a decisive advantage. It took a long time before a second look was taken. Only when McNamara's system analysts came to the Pentagon was it discovered that the Soviet divisions' strength was not only qualitatively lower (this had been widely assumed before, even though the extent of qualitative advantage was uncertain) but even more importantly, Soviet and Warsaw Pact divisions were smaller in the number of troops per division. It turned out that the Soviets had two million men at arms vis-à-vis the 960,000 of the United States, but with the forces of the allies in Europe taken into account the Warsaw pact was in fact outnumbered by NATO with only four and half million troops against the six million in the NATO armies.[37] In fact, neither side could expect an easy victory in general conventional war.

Apart from the global conventional balance, two particular theater balances should be considered. In Cuba, the USSR deployed not only the missiles but also a substantial contingent of Soviet troops to support and protect the missile deployment. Air defense was provided by a dense network of SA-2 missiles. The Soviets also dispatched a regiment of 40 MIG-21 fighters, 33 Il-28 light bombers, FROG rockets, and some 42,000

Soviet troops. The deployment to Cuba was the largest amphibious operation in Soviet history.[38] Only a part of that story was known to the United States. U.S. intelligence largely underestimated the strength of the Soviet forces. Yet, Washington assembled a formidable force for the possible invasion of Cuba. One hundred thousand Army and 40,000 Marine combat troops were readied in Florida; the invasion plans involved 579 tactical aircraft and 183 ships including eight aircraft carriers.[39] With its advantage in numbers and short lines of communication, the United States would certainly have prevailed in a conventional battle in Cuba. But the analogous situation with a reversed outcome would be valid for another Cold War hotspot, West Berlin. This showcase of the free world, deep inside communist East Germany, was militarily indefensible. There was no chance of stopping a determined Soviet grab of Berlin; the only option would be escalation and eventually retaking the city after at least a year devoted to mobilization.[40]

In terms of technological advantage, the U.S. enjoyed some limited edge since, in general, the Soviet technologies were considered to be lagging behind.[41] However, in most combat systems the Soviet systems' inferiority would not be likely to be decisive. Two major combat systems – fighter jets and tanks – can be used to illustrate the general situation. The USSR could answer to the U.S.'s newest fighters, Navy's F-4 Phantoms (in service from 1960) and USAF's F-105 Thunderchiefs (in service from 1958), with its own new agile MIG-21s (in service from 1959). To the U.S.'s newest M-60 tanks (in service from 1960) there were new T-62s, entering service since 1961. Certainly the United States deployed some technologies that were largely unavailable to the Soviet Union. This was particularly true in naval warfare where the Soviets had no aircraft carriers as the Soviet navy's bet was on submarines. But, in general, the U.S. advantage in technology cannot be considered decisive.

Certainly the U.S.'s technological advantage was not enough to provide full information on Soviet targets. This was particularly true of Cuba. In Fall 1962, the U.S. reconnaissance satellites were regularly in duty over the Soviet Union, but not over Cuba to spot the USSR buildup in full.[42] After Eisenhower's decision to break relations with Cuba, United States lost its embassy in Havana which served as a first-hand listening post. Thus, on the spot, Washington had to rely on a diminishing number of CIA informants, deep-cover agents, and notoriously unreliable exiles.[43] The cost was imperfect intelligence. The United States spotted the Soviet military build-up in Cuba but, until U-2 photographs were taken in mid-October, Kennedy and his advisors had no idea of its scope or magnitude. Even then, the U.S. underestimated the strength of the Soviet forces in Cuba by four to 10 times.[44] Considering the importance that the administration attached to Cuba, its inability to get accurate intelligence is more than remarkable.

Not only was the conflict over Soviet missiles in the Caribbean a part of the Cold War, but also for the Americans it was an unprecedented intrusion into the region where the U.S. had claimed dominance more than a hundred years ago. As Thomas Paterson observed,

> President Kennedy spent as much or more time on Cuba as on any other foreign policy problem. Cuba stood at the center of his Administration's admitted greatest failure, the Bay of Pigs, and its alleged greatest success, the missile crisis.[45]

After the Bay of Pigs, the administration started an aggressive covert operation, Mongoose, aimed at overthrowing Castro's regime. Being part of the central Cold War confrontation between the United States and the Soviet Union, in a region where the United States was oversensitive, the centrality of conflict in this case is beyond any doubt.

Turning to perceptional factors, how did the United States perceive its enemies in the crisis? Certainly Washington had a substantial history of hostility toward the Soviet Union that dated way back to 1917. The hostility was interrupted by a brief period of Second World War cooperation which abruptly ended with the onset of the Cold War. With Cuba, the United States' position was different in the duration of hostility, yet its intensity was arguably even stronger. At least, the Soviets by 1962 often appeared as acceptable adversaries who could be talked with. With Castro, this was rare as almost all official contacts were broken.

Contrary to Cuba, the United States and the Soviet Union enjoyed a wide range of various contacts all over the world. Both leaders had a number of formal and informal communication channels available. They even had a chance to meet each other in Vienna in 1961. Even the beginning and escalation of the Cold War did not interrupt the formal diplomatic relations that had lasted since U.S. recognition of the USSR in 1933. Thus, official communication via the U.S. embassy in Moscow and, importantly during the crisis, via the Soviet embassy in Washington was available. The embassy in Washington also provided informal communication channels. There, KGB agent Georgi Bolschakov was used by both Kennedy and Khrushchev to handle sensitive communication.[46] Also Soviet Ambassador Dobrynin himself informally met twice with Robert Kennedy. Their second meeting allowed transferring the message about the secret deal on Jupiters that substantially helped to defuse the crisis.

Fairly frequent communication cultivated the U.S. perception of Soviet rationality. Khrushchev's previous track record showed a man prone to risk taking, perhaps irresponsible, but not irrational. During the crisis, Kennedy developed a new understanding of the Soviet leader, one of a leader "who had bungled into the crisis" and wanted to find a way out without losing face.[47] Thus, importantly, he wanted Khrushchev to react calmly and he believed that it was possible.[48] Nevertheless, there was a great fear of irrationality in Washington. The roots of this feeling, however, did not aim directly at the Moscow leadership as if it was somehow exceptionally irrational. Rather the uncertainty lay in the Soviet command and control system, making the Americans afraid that the decision to launch missiles could be made by some low-level Soviet commander in charge of a missile battery in Cuba.[49] This probably helped to strengthen deterrence in a way that Thomas Schelling describes as a threat that leaves something to chance.[50] It was particularly relevant as a considerable number of ExCom members questioned Soviet readiness to respond to a U.S. airstrike.

In fact the perception of Soviet resolve varied in the ExCom. Hawks like Dillon apparently believed that a harsh Soviet reaction could be avoided with a quick strike and "a statement at the time saying this is all there is to it."[51] He may have been right. When Soviet decision makers discussed in October how they should react if the U.S. attacked the missiles, this turned out to be one of the likely options. The discussions in Moscow largely covered the possibility of acting against Berlin or the Turkish missiles, but the fear of escalation among the Soviet leadership was strongly making no response equally possible.[52] However, this was not a dominant feeling in the ExCom. The President was gravely concerned about the USSR's perception of his own resolve, not questioning Khrushchev's. Most ExCom members were convinced that the Soviets would retaliate in some way. George Ball summed it up during one of the first meetings: "You go in there with a surprise attack. You put out all the missiles. This isn't the end. This is the beginning."[53] McNamara and Robert Kennedy were also strongly of this opinion, objecting that the Soviets would certainly take some action, if not in Cuba then in Berlin or Turkey.[54] Not unusually, the President's view was not dissimilar to his brother's.

International factors other than third-party deterrence were also influencing the decision making in Washington. Unsurprisingly, no other state could muster a military power equal to the two superpowers involved in the crisis, so third-party deterrence was out of the question. Yet the looming Cold War often made the importance the two superpowers attached to the opinion of third countries far outweigh their real importance. Alliance politics and international action legitimacy certainly mattered.

The prevailing perception in Washington was that European allies would be uneasy with the airstrike. The Europeans had few reasons to think that the missile in Cuba could constitute a "casus belli." They had been living with hundreds of Soviet missiles next door. The President colorfully described the European allies who were accustomed to living with Soviet missiles as thinking of the U.S. as "fixated ... slightly demented on this subject."[55] Destroying the missiles but losing Berlin in exchange looked unacceptable to the allies. A letter from British Prime Minister Macmillan was clear on that.[56] Important in choosing a blockade, not an airstrike, it allowed for a gradual response and maximum allied support. This was not Washington's only sensitivity to its allies. When Khrushchev requested the missile trade, Washington had to publicly refuse, even though the Jupiters were of no military value. It was rightly believed that the allies would not be happy about a deal that would appear as trading U.S. security for their own. Complicating things further, the Jupiters were deployed under a dual-key arrangement and the missiles formally belonged to Turkey. Both with the missile trade and with the blockade, the allied position influenced the decision making in Washington.

Allied support was important for enhancing action legitimacy. In this respect, the Organization of American States was instrumental. But also of great importance was the normative belief shared by a large number of ExCom members about the moral unacceptability of a surprise attack. It was George Ball who brought into the discussion the comparison of the surprise airstrike on Soviet missiles with Pearl Harbor, the Japanese attack that the Americans saw as a sneaky, immoral action.[57] Even the hawkish ExCom members like Dillon and McCone were uncomfortable with America acting like Japan at Pearl Harbor. Robert Kennedy was also of this opinion and he could not imagine how the President would justify such an action, as he observed "for 175 years we had not been that sort of country."[58]

Speaking about another normative issue of interest – the nuclear taboo – it can be safely noted that U.S. decision makers engaged in few serious decisions about using nuclear weapons first.[59] The nuclear first strike option was not debated. It was argued that nuclear weapons should not be used against Cuba though it might be necessary to use them in a compensatory attack against the USSR if the missiles were launched from Cuba.[60] But likely deterrence and the availability of conventional alternatives to strike Cuba were at the time more important than the norm itself.[61] In fact, it was the crisis itself, the moment when the world believed it was close to the brink of nuclear war, which later played a formative role in the institutionalization of the taboo.

Contrary to international sensitivities, the Kennedy administration would have faced little constraint domestically if it had decided to take the Soviet missiles out with an airstrike. In fact Kennedy was pressed on Cuba by domestic politics. Congressional elections were scheduled for November and the President's policy on Cuba, with the Bay of Pigs and its aftermath, earned him a poor approval rating.[62] Appearing weak on Castro or communists was unpopular. By September, when it was recognized that Soviet deliveries and troops were flowing into Cuba, but before the missiles were discovered, some 70 percent of Americans wanted Kennedy to take some sort of action on Cuba.[63] In October, when the President made his decision and announced the naval blockade, the Congress led by Senator Russell and Senator Fulbright responded strongly in favor of an airstrike.[64] The President had good reasons to resist, but domestic politics was not one of them. He had well recognized that the air strike "would be a swifter and more popular means of removing the missiles before election day."[65]

Should he have decided to do so, he would have faced opposition from some of his top advisors. Together with other ExCom members, two of his most influential aides, Secretaries McNamara and Rusk, did not favor the airstrike. This was also the position of the President's brother; however, Robert was sensitive to John's position and certainly if the President had decided for the airstrike he would have consented. Furthermore, even McNamara and Rusk largely opposed the airstrike at that moment, but did not rule it out for the future when other options were exhausted. True, McNamara argued during the October 16 ExCom meeting that the strike would be impossible once the missiles became operational, but later on October 27 he declared that "we must now be ready to attack Cuba.... Invasion had become almost inevitable."[66] Similarly to the others, his position was influenced by last-resort feelings.

Notably this last-resort feeling was strong among the early supporters of the airstrike. It was argued that the airstrike would be difficult, risky, and perhaps impossible when the missiles in Cuba became operational. Waiting was only to the advantage of the Soviets. The President himself did not believe that the last resort moment had come. First, there was the option of enlarging the blockade, particularly to include petroleum, oil, and lubricants, a preferable option for McNamara and Bundy.[67] Even if the stronger blockade failed to push the Soviet missiles out of Cuba, Kennedy would have at least one option short of the airstrike reserved. Secretary Rusk would use former UN official Andrew Cordier to arrange a public appeal from UN Secretary General U Thant for a missile trade. This would reduce the allies' opposition to such a trade and Kennedy would then be able to accept.[68]

The last thing to consider before the effects of the deterrence threat is the regime type. It would be a lazy practice, but at this point the appropriate paragraph from the first empirical chapter can be safely copy-pasted. To sum it up, the U.S. were a full democracy with firm civilian control of the military, which Secretary McNamara had imposed over services to their little liking. Nothing surprising; the most interesting part is to follow.

Deterrence worked in the Cuban Missile Crisis. It is well-known that peace was endangered by accidents and unintended escalatory events like the shooting down of the U.S. U-2 over Cuba by Soviet forces which were explicitly ordered by Khrushchev not to do so. No one knows what could have happened if such uncontrolled and unintended escalation had continued. But accidents notwithstanding, deterrence worked and importantly for this study, the threat of nuclear retaliation contributed to its working. In Washington, the top decision makers did not believe that the U.S. could face war with confidence, even with its vast nuclear power.[69] General Taylor could not offer more than a 90 percent success rate for a strike against the missiles at Cuba, and the option of a first strike against the Soviet Union was not even discussed. The President, McNamara and others were afraid of escalation which could lead to a nuclear confrontation. Even though the Soviets would gain little by retaliating, it was uncertain whether some decision to retaliate with nuclear weapons could be taken by a local commander of a missile battery in Cuba.

Equally important was the threat of Soviet conventional retaliation. A reciprocal attack against the U.S. missiles in Turkey or some action against vulnerable West Berlin was believed to be likely. Robert Kennedy and McNamara strongly voiced those fears. Robert believed that the Russians would have to react somehow, if not in Cuba then certainly in Berlin.[70] Secretary McNamara thought about other places as well, assuming that the U.S. action against Cuba would probably lead to a Soviet reaction "some place in the world," likely in Berlin, or against Turkey, possibly even Iran and Korea.[71] The fear did not appear unwarranted. Not only was such a reaction debated in Moscow, but the Warsaw Pact's armed forces announced an alert.[72] Some of the Soviet moves could be challenged, but militarily there would be no chance to save West Berlin from a Soviet grab.

Interestingly, denial strategies worked less effectively then retaliatory ones. Their full effectiveness was largely mitigated by holes in U.S. intelligence about the Soviet contingent in Cuba. Washington underestimated the strength of the Soviet forces by a factor of four to 10 and was thus unable to grasp all the likely costs of attacking Cuba. Likely, even had the intelligence known the full scale of Soviet deployment, the airstrike costs would have been acceptable to the U.S. Some 24 SA-2 sites with 144 missiles protected the Soviet nuclear missiles in Cuba together with a regiment of MIG-21; a strong but not impenetrable defense. Yet the U.S. was also getting ready for the need of invasion, which the military assumed would cost 18,500 American lives in 10 days of combat.[73] The estimate would have certainly been higher had the U.S. known that there were more than 40,000 Soviet soldiers in Cuba. Whether this would have altered the effectiveness of conventional denial is uncertain. Certainly, with the information available to Washington, conventional denial was not considered prohibitive to an action against Cuba.

Neither was nuclear denial. Again, there was an imperfection in U.S. intelligence in Cuba. The Soviets deployed nuclear-armed FROG missiles and cruise missiles to support their contingent. Whereas U.S. intelligence discovered the FROGs, it assumed that they were not nuclear-armed.[74] The full extent of the Soviet tactical nuclear force in Cuba was unknown to the U.S. invasion planners and could hardly serve as a meaningful deterrent.[75]

What is the main summary of this chapter that serves this book as a control for a quality of asymmetry? Most importantly, it is that quality matters. The Soviets were believed to have a fairly large, though unsophisticated, nuclear arsenal. While the U.S. military could offer the President a high probability of knocking it out they had to concede they would probably miss some targets. Thereby the Soviets enjoyed at least some limited second strike capability. Their retaliation was anything like assured, but it was also far from implausible. The leaders in Washington were sensitive to this threat and reluctant to risk the nuclear war. Yet, it is notable that the existing fear of nuclear retaliation was clearly accompanied by the equal fear of Soviet conventional retaliation, which looked likely and credible given the perceived Soviet preponderance in Europe in general and against West Berlin in particular. Thus, despite being gravely concerned by the Soviet deployment in Cuba, the Kennedy administration opted for less confrontational strategies and war was avoided. All this, of course, happened in interplay with other factors, and at least some deserve to be noted in this summary. In particular, the Pearl Harbor style surprise attack's illegitimacy and the concerns over allied reaction occupied the minds of the U.S. decision makers. And last but not least, there was a surprisingly favorable perception of the Soviet leadership's rationality in Kennedy's eyes – Moscow could be dealt with, spoken with, and, as it was a rather rational adversary, it was quite possible to live in peace with her.

Table 6.1 U.S.–USSR, 1962

Concept	Description	Values
The deterrer's nuclear arsenal	Key physical attributes of respective nuclear complexes; weapon types; numbers; delivery vehicles; command, control, and communication systems	350+ (40 ICBMs, 60 SLBMs, 160 strategic bombers, all rather unsophisticated; hundreds of sub-strategic weapons; primitive early warning and command and control)
Nuclear asymmetry	Situation when a substantially smaller (quantitatively) and less sophisticated nuclear force (qualitatively) faces a qualitatively and quantitatively larger nuclear force	Strong asymmetry (quantitative superiority by a factor of approx. 10 in strategic weapons; substantial qualitative edge)
Second-strike criterion	Nuclear posture that has the ability to survive enemy attack, make and communicate decision to retaliate, overcome enemy's active defense, and destroy a valuable target despite its passive defense	Limited (uncertain but likely survivability of some strategic forces; good survivability of sub-strategic forces; adequate delivery vehicles)
General conventional preponderance	Situation when one side's armed forces are in general substantially stronger in terms of numbers, technology, training, and employment strategies	Rough balance (the challenger's advantage in navy and air force; the deterrer's advantage in land force)
Theater conventional preponderance	Situation when one side's armed forces are substantially stronger in terms of numbers, technology, training, and employment strategies on a theater where targets valuable to the challenger can be found	Deterrer (Strong and unchallengeable advantage against Berlin)

continued

Table 6.1 Continued

Concept	Description	Values
The challenger's technological advantage	Significant advantage of the challenger's major weapons systems that would be employed in case of conflict in terms of state-of-the-art sophistication over the deterrer's weapons systems	Limited (some edge in certain weapons platforms)
Availability of information	The challenger's knowledge about targets' location and defensive systems, which is established from sources that the challenger deems credible	Limited (holes in intelligence about USSR deployment in Cuba)
Centrality of conflict	Absolute importance of the conflict dyad between the challenger and the deterrer from the challenger's subjective perspective and its relative importance to other existing conflict dyads where the challenger is a party	Yes (part of cold war; in a sensitive region for the U.S.; administration's special sensitivity after Bay of Pigs)
Perceived resolve of the deterrer	The challenger's perception of the deterrer's commitment to fight should the deterrence fail	Strong (but questioned by hawks)
Institutionalization of mutual relations	Degree of shared expectations about the requirements of stable deterrence and the existence of proven formal or informal communication channels	Strong (formal and informal communication channels available; history of frequent contacts)
History of hostility	Track record of conflict in the dyad that shapes the understanding of other side's intentions	Strong (cold war)
Last resort	Situation when the challenger sees only the options to strike, or to live up to the development he tries to prevent	Limited (UN-arranged missile trade as back-up option; last-resortness seen by proponents of strike)

Perception of the deterrer's rationality	The challenger's perception of the deterrer's rationality, particularly whether the challenger believes it possible to live with the nuclear-armed deterrer in the long term	Yes (leadership largely viewed as rational; fears over irrational decisions of field commanders)
Third-party deterrence	Threat of military involvement into the original conflict by a third party, most likely the allies of the original deterrer, that decisively influenced the challenger's decision	None (no one had the power to deter superpowers)
Alliance politics	Sensitivity of the challenger to possible impact of his action on the relations with his allies	Strong (NATO opposition to strike and missile deal)
International action legitimacy	The challenger's sensitivity to the international normative expectation of non-intervention	Strong (fears over Pearl Harbor stigma; legitimacy seeking actions, e.g. in the OAS)
Nuclear taboo	Normative prohibition on the use of nuclear weapons	Emerging (nuclear use not even discussed, but largely driven by the logic of consequences)
Domestic action legitimacy	Level of support for military solution among the challenger's population	Strong (domestic audience more hawkish than administration)
Opposition among influential decision makers	Negative view on military solution by influential part of the challenger's government	Strong (including influential McNamara and Rusk, but depending on lack of last-resortness)
Regime type	The challenger's position on the democracy-nondemocracy axis and on the militarized-nonmilitarized axis	Democracy, strong civilian control of armed forces

continued

Table 6.1 Continued

Concept	Description	Values
Nuclear retaliation	Threat that the deterrer will use its nuclear weapons against targets valuable to the challenger, except of targets that are directly related to pursuit of the challenger's objectives	Strong (President, McNamara, and others afraid of escalation leading to nuclear war)
Nuclear denial	Threat that the deterrer will use its nuclear weapons against targets that are directly related to pursuit of the challenger's objectives in order to prevent him from attaining the objectives, or in order to make it unacceptably costly	Limited (Soviet tactical nuclear options in Cuba not known)
Conventional denial	Threat that the deterrer will use its conventional weapons against targets that are directly related to pursuit of the challenger's objectives in order to prevent him from attaining the objectives, or in order to make it unacceptably costly	Limited (Soviet strength in Cuba underestimated, but would not likely be considered prohibitive)
Conventional retaliation	Threat that the deterrer will use its conventional weapons against targets valuable to the challenger, except of targets that are directly related to pursuit of the challenger's objectives	Strong (strong concerns about Soviet action against West Berlin or Turkey)

Notes

1 John Lewis Gaddis, *We Now Know: Rethinking Cold War History* (Oxford: Oxford University Press, 1997) 260.
2 Richard Ned Lebow and Janice Gross Stein, *We All Lost the Cold War* (Princeton: Princeton University Press, 1995) 67–69.
3 Lawrence Freedman, *Kennedy's Wars: Berlin, Cuba, Laos, and Vietnam* (Cary: Oxford University Press, 2000) 170.
4 Freedman, *Kennedy's Wars*, 163–164.
5 Ned Lebow and Gross Stein, *We All Lost the Cold War*, 99.
6 John Lewis Gaddis, *The Cold War: A New History* (New York: Penguin Books, 2005) 75–76.
7 Theodore C. Sorensen, *Kennedy* (New York: Harper & Row, 1965) 670; Ned Lebow and Gross Stein, *We All Lost the Cold War*, 95–102.
8 Marc Trachtenberg, "The Influence of Nuclear Weapons in the Cuban Missile Crisis," *International Security*, 10/1 (Summer 1985) 141.
9 Sheldon M. Stern, *The Week the World Stood Still: Inside the Secret Cuban Missile Crisis* (Stanford: Stanford University Press, 2005) 41.
10 Freedman, *Kennedy's Wars*, 177.
11 Stern, *The Week the World Stood Still*, 45–54.
12 Freedman, *Kennedy's Wars*, 178–181.
13 Freedman, *Kennedy's Wars*, 182–187.
14 Stern, *The Week the World Stood Still*, 72–74.
15 Ned Lebow and Gross Stein, *We All Lost the Cold War*, 118.
16 Freedman, *Kennedy's Wars*, 195.
17 For details see Curtis A. Utz, *Cordon of Steel: The U.S. Navy and the Cuban Missile Crisis* (Washington: Naval Historical Center, 1993).
18 Freedman, *Kennedy's Wars*, 209.
19 Gaddis, *We Now Know*, 276.
20 David A. Welch and James G. Blight, "An Introduction to the ExComm Transcripts," *International Security*, 12/3 (Winter 1987–1988) 5–29; for details on the missiles in Turkey, see Philip Nash, *The Other Missiles of October: Eisenhower, Kennedy, and the Jupiters, 1957–1963* (Chapel Hill: University of North Karolina Press, 1997).
21 Freedman, *Kennedy's Wars*, 216; Ned Lebow and Gross Stein, *We All Lost the Cold War*, 120.
22 Gaddis, *We Now Know*, 276.
23 Ned Lebow and Gross Stein, *We All Lost the Cold War*, 137.
24 Gaddis, *We Now Know*, 278.
25 Vladislav M. Zubok, *A Failed Empire: The Soviet Union in the Cold War from Stalin to Gorbachev* (Chapel Hill: The University of North Carolina Press, 2007) 131.
26 Robert S. Norris and Hans M. Kristensen, "The Cuban Missile Crisis: A Nuclear Order of Battle, October and November 1962," *Bulletin of Atomic Scientists*, 68/6 (November 2012) 87.
27 Freedman, *Kennedy's Wars*, 172.
28 Norris and Kristensen, "The Cuban Missile Crisis," 87.
29 The number is based on Stephen J. Cimbala, *Nuclear Deterrence in the 21st Century* (Westport: Greenwood Press, 2000) 49. Cimbala claims that the Soviets also deployed four newer SS-N-5s, but most sources suggest that these missiles were only available from 1963.

30 Norris and Kristensen, "The Cuban Missile Crisis," 87.
31 Trachtenberg, "The Influence of Nuclear Weapons in the Cuban Missile Crisis," 156–161.
32 Norris and Kristensen, "The Cuban Missile Crisis," 87.
33 Gareth Porter, *Perils of Dominance: Imbalances of Power and the Road to Vietnam* (Berkeley: University of California Press, 2005) 15.
34 Cimbala, *Nuclear Deterrence in the 21st Century*, 53.
35 ExCom Meetings October 1962, "White House Tapes and Minutes of the Cuban Missile Crisis," *International Security*, 10/1 (Summer 1985) 175.
36 Freedman, *Kennedy's Wars*, 216.
37 Gaddis, *Strategies of Containment*, 205–206.
38 Gaddis, *We Now Know*, 227; Ned Lebow and Gross Stein, *We All Lost the Cold War*, 76.
39 Freedman, *Kennedy's Wars*, 175.
40 Freedman, *Kennedy's Wars*, 104.
41 Porter, *Perils of Dominance*, 3.
42 Freedman, *Kennedy's Wars*, 167
43 Thomas G. Paterson, *Kennedy's Quest for Victory: American Foreign Policy, 1961–1963* (Cary: Oxford University Press, 1989) 130.
44 Gaddis, *We Now Know*, 267.
45 Paterson, *Kennedy's Quest for Victory*, 123.
46 Ned Lebow and Gross Stein, *We All Lost the Cold War*, 68.
47 Ned Lebow and Gross Stein, *We All Lost the Cold War*, 145.
48 Freedman, *Kennedy's Wars*, 195.
49 Trachtenberg, "The Influence of Nuclear Weapons in the Cuban Missile Crisis," 154.
50 Thomas C. Schelling, *Arms and Influence* (New Haven: Yale University Press, 1966) 94–97.
51 Freedman, *Kennedy's Wars*, 177, 189.
52 Ned Lebow and Gross Stein, *We All Lost the Cold War*, 137.
53 Freedman, *Kennedy's Wars*, 178.
54 Freedman, *Kennedy's Wars*, 180; Trachtenberg, "The Influence of Nuclear Weapons in the Cuban Missile Crisis," 142.
55 Freedman, *Kennedy's Wars*, 179.
56 Freedman, *Kennedy's Wars*, 190.
57 Stern, *The Week the World Stood Still*, 53.
58 Freedman, *Kennedy's Wars*, 176–187.
59 Tannenwald, *The Nuclear Taboo*, 6.
60 Freedman, *Kennedy's Wars*, 191.
61 Tannenwald, *The Nuclear Taboo*, 253.
62 Paterson, *Kennedy's Quest for Victory*, 126.
63 Freedman, *Kennedy's Wars*, 161.
64 Freedman, *Kennedy's Wars*, 194.
65 Ned Lebow and Gross Stein, *We All Lost the Cold War*, 95.
66 Trachtenberg, "The Influence of Nuclear Weapons in the Cuban Missile Crisis," 142.
67 Ned Lebow and Gross Stein, *We All Lost the Cold War*, 128.
68 Gaddis, *We Now Know*, 267; Ned Lebow and Gross Stein, *We All Lost the Cold War*, 127–128.
69 Trachtenberg, "The Influence of Nuclear Weapons in the Cuban Missile Crisis," 147.

70 Freedman, *Kennedy's Wars*, 180.
71 Trachtenberg, "The Influence of Nuclear Weapons in the Cuban Missile Crisis," 142.
72 Ned Lebow and Gross Stein, *We All Lost the Cold War*, 116.
73 Freedman, *Kennedy's Wars*, 175.
74 Gaddis, *We Now Know*, 275.
75 Cimbala, *Nuclear Strategy in the 21st Century*, 54, fn. 21.

7   Putting the pieces together

Individually, each case represents a fascinating story of historical uniqueness. Yet taken together they allow something even more fascinating, at least to students of deterrence. They provide an imperfect, perhaps largely preliminary, but for the moment probably the best available, opportunity to distinguish historical uniqueness from the regularities of deterrence in small-to-large dyads. The observations that rest in the study of five selected cases are powerful and important. This is true particularly with respect to the voluminous yet empirically underdeveloped literature biased by the analytical primacy of nuclear deterrence. There is no ground for such a bias.

Key findings

The common expectation of mainstream deterrence theory is that in the realm of nuclear deterrence, or more accurately in the realm of nuclear retaliation, the possible costs of aggression will always outweigh its benefits. The idea originated from the understandable proposition that no benefits will warrant putting national survival in jeopardy,[1] but was further broadened to the point that even a single nuclear explosion over a major city of the challenger would probably do the job as well. Allegedly this is not the case with conventional weapons. On the contrary, many potential adversaries are believed to be undeterrable with conventional weapons. Deterring those who can be deterred with conventional weapons requires denying them the expectation of a quick military victory, and, should the deterrence fail, the threat of imposing suffering and destruction [retaliation] is less effective than the threat to defeat the adversary and deny his objectives.[2] To summarize conventional wisdom, conventional deterrence works as a denial strategy. Yet conventional denial is not unproblematic. Richard Betts argues that "confidence in conventional deterrence by defense is unwarranted because denial is an inherently weaker deterrent than punishment – the costs of failure are much smaller."[3] Betts is somewhat sympathetic to Huntington's concept of conventional retaliation yet finds it unfeasible for operational reasons and for the destabilizing effects it allegedly has. Regrettably this discussion about conventional deterrence from the 1980s was terminated by the end of Cold War. The prevailing option thus remains that the threat of awful nuclear retaliation is not only more stabilizing than any conventional threat as the latter is inherently more prone to miscalculation, but also that nuclear deterrence deserves primacy in scholarly attention. Few theorists have considered that conventional threats can be more robust than nuclear ones.

But where conventional and nuclear weapons exist alongside each other – and it is hard to imagine that a nuclear-armed state would not be armed conventionally as well – both conventional and nuclear deterrence operate. In fact, conventional deterrence can be the more influential of the two. This is particularly true with a small nuclear arsenal. Revealingly, the U.S. ambassador in Korea and the commander of U.S. forces at the theater in 1994 summed up their shared feeling in asking: "why are we going to risk killing a million people? A bomb or two can't even do that."[4] Similar patterns occurred in the U.S.–China case and the Soviet Union-China case. In all three cases, conventional threats clearly outweighed nuclear ones. In 1994, Ambassador Laney and General Luck, as well as top decision makers in the Clinton administration, were concerned about North Korea's artillery pieces targeting Seoul, but not about the possibility of the DPRK's one or two crude and possibly nonexistent bombs being used against the United States or its allies. Contemplating the possibility of destroying China's emerging nuclear arsenal, Kennedy's Joint Chiefs were not concerned that some Chinese nuclear weapons were already ready to use, or that some had possibly been supplied by the Soviet Union, even though both were possible. Rather their minds were preoccupied with the possibility of China's covert or overt retaliatory aggression against U.S. allies in Asia. Similarly, empirical evidence does not support the claim that some 50 vulnerable Chinese nuclear bombs with imperfect delivery systems deterred the Soviets in 1969. Moscow was rather concerned by the threat of conventional war at a distant theater where the USSR was precariously weak, and it correspondingly reacted by a gradual and massive buildup of its conventional forces in the region. Of course, this is not to say that nuclear threats are necessarily weak or lack credibility, while conventional threats are strong and credible. Neither is necessarily true. But under circumstances both can be. Scholars cannot avoid paying attention to the role conventional weapons play in nuclear dyads. It may be true that conventional deterrence is, in general, more prone to miscalculation – perhaps to most leaders the costs of conventional conflict may appear more acceptable[5] – but that does not warrant assuming that its effects are negligible in nuclear dyads. In fact, all the cases showed at least comparable effects of conventional threats to nuclear threats in the challenger's decision making. The three cases which are discussed in the previous paragraph highlight that conventional deterrence can be stronger than nuclear deterrence. In the two other cases, the effects were roughly equal. Conventional deterrence that succeeded in the previous cases failed in Osiraq because Iraq lacked strong enough means of punishing Israel for its action, not least because its forces were tied up in a war with Iran. Iraq's nuclear deterrence did not perform better. In the Cuban Missile Crisis, Kennedy's administration was clearly afraid of escalating the conflict to the level of full-scale war, which would inevitably be nuclear. However, a comparable restraint was put on U.S. decision makers by the threat of Soviet conventional action against West Berlin.

Having this in mind, it is clear that students of deterrence should pay more attention to its conventional dimension. It is striking that John Mearsheimer's seminal book on conventional deterrence is now more than three decades old, yet no one has written anything comparable. Second, students of nuclear deterrence, even if by definition not interested in conventional deterrence, must carefully control for the effects of the latter in nuclear dyads. This would not make their already difficult job easier; quite to the contrary. But it should greatly improve the validity of their results.

The other observation about conventional deterrence is at least equally important. It appears to be the single most important underpinning for a departure from the analytical primacy of nuclear deterrence. So far, conventional strategies have been largely identified as denial strategies.[6] However, an empirical record shows that conventional retaliation constitutes an unduly omitted phenomenon. The threat of conventional retaliation is the best plausible explanation in three out of the four deterrence successes studied. The Americans were truly afraid of China's action against their allies when considering taking out Beijing's emerging nuclear program. Previously, the PRC had shown its ability to act militarily and decisively in its proximity, most profoundly in the Korean War. Strategically, Beijing had an enormous advantage in manpower available in the theater and could use it against U.S. allies such as South Korea, South Vietnam, Taiwan or others. When the Soviets considered action against China's nuclear arsenal they found their hands tied by the threat of a Chinese onslaught on the Soviet Far East. The Chinese enjoyed an advantage of more than two to one in manpower and substantially shorter and less vulnerable lines of communication. Moscow responded with the major redeployment of its conventional troops, but this was only finished several years later. Furthermore, Moscow knew that even if the Soviet Far East could be defended, and perhaps China could even be conventionally defeated in the battlefield, this would be enormously costly. Many years later, President Clinton was constrained in his dealing with North Korea's nuclear crisis by the threat of artillery barrages and, possibly, the DPRK's armored sweeps against the vulnerable capital of his South Korean ally. Again, while little to no serious considerations were taken with respect to the possibility of the DPRK's nuclear retaliation, the threat of a new Korean War occupied the minds of U.S. decision makers, and major troop reinforcements as well as major boosts of counterbattery capabilities were being prepared. Moreover, not only did the threat of conventional retaliation play a decisive role when the deterrer's nuclear arsenal was vulnerable or virtually nonexistent, it also substantially supported the Soviet nuclear threat in constraining the U.S. ability to take out Soviet missiles in Cuba by an airstrike in the Cold War's most serious crisis. There, the Soviet Union's ability to take action against Berlin was instrumental.

This runs contrary to common understandings and has profound implications for theory and strategy. The standard expectation about conventional retaliation is that it rarely succeeds. States are allegedly willing to accept fairly high costs including the enemy's punishment of its civilian population; conventional weapons supposedly inflict only limited damage; modern states can minimize their vulnerability to conventional attacks; and not only leaders but also the civilians who are being punished are ready to accept this punishment as an acceptable exchange for the expected benefits.[7] Yet, these observations about conventional deterrence are mostly drawn from wartime cases of strategic bombing. Neither Germany, nor Japan surrendered to conventional Second World War bombing. Vietnam and Korea are similar cases. Yet, apparently these cases have only limited validity for prewar deterrence, or at least that is what the empirical evidence in this study suggests. The theory must be reconsidered in its understanding of conventional retaliation and pay more attention to prewar threats of selective conventional retaliation. It is not correct to keep identifying successful conventional deterrence with conventional denial and conventional retaliation with victory in an all-out war; nor is conventional retaliation necessarily ineffective.

But theory is not alone in omitting conventional retaliation. While empirical evidence shows the remarkable success of conventional retaliatory threats, such threats are rarely a common part of national strategies, or at least they are seldom explicitly articulated. Usually, in the conventional weapons realm, states threaten to defend themselves (denial) or, if they feel strong enough, they also add the threat of fighting and wining a conventional war. The threat of selective conventional retaliation is largely omitted from the national strategies of most countries, even though it can apparently replace nuclear threats, at least for some contingencies.

Empirical evidence also shows remarkably well which single factor is critical for a robust threat of conventional retaliation. It is possible to state with a high degree of confidence that theater conventional preponderance makes a difference. This runs against common expectations about conventional deterrence. There, the general theoretical assumption is that deterrence will hold where defense is stronger than offense.[8] For denial strategies which tend to be overwhelmingly equaled with conventional deterrence, this is probably true. Yet my results convincingly demonstrate that, for successful

Table 7.1 Summary of empirical results

	U.S.–China	USSR–China	Israel–Iraq	U.S.–DPRK	U.S.–USSR
The deterrer's nuclear arsenal	0–5	Approx. 50	0–1	0–2	350+
Nuclear asymmetry	Strong asymmetry	Strong asymmetry	Strong asymmetry	Strong asymmetry	Strong asymmetry
Second-strike criterion	None	None	None	None	Limited
General conventional preponderance	Challenger	Challenger	Challenger	Challenger	Rough balance
Theater conventional preponderance	Deterrer	Deterrer	Challenger	Deterrer	Deterrer
The challenger's technological advantage	Strong	Strong	Strong	Strong	Limited
Availability of information	Limited	Strong	Strong	Limited	Limited
Centrality of conflict	Yes	Yes	Yes	Yes	Yes
Perceived resolve of the deterrer	Strong	Strong	Strong	Strong	Strong
Institutionalization of mutual relations	Limited	Strong	Limited	Limited	Strong
History of hostility	Strong	Limited	Strong	Strong	Strong

Last resort	Strong/limited	Limited	Strong	Limited	Limited
Perception of the deterrer's rationality	No	No	No	No	Yes
Third-party deterrence	Limited	Limited	Limited	Limited	No
Alliance politics	No	No	Strong	Strong	Strong
International action legitimacy	Limited	Limited	Strong	Limited	Strong
Nuclear taboo	Emerging	Emerging	At least emerging	Strong	Emerging
Domestic action legitimacy	Strong	Likely not applicable	Strong	Strong	Strong
Opposition among influential decision makers	Strong	Strong	Strong	Limited	Strong
Regime type	Democracy, strong civilian control of armed forces	Non-democracy, strong party control of armed forces	Democracy, civilian control of armed forces	Democracy, strong civilian control of armed forces	Democracy, strong civilian control of armed forces
Nuclear retaliation	Limited	Limited	Limited	Limited	Strong
Nuclear denial	Limited	Limited	Limited	Limited	Limited
Conventional denial	Limited	Limited	Limited	Limited	Limited
Conventional retaliation	Strong	Strong	Limited	Strong	Strong

conventional retaliation, the requirements may be quite opposite. The local advantage of offense over defense can constitute good deterrence. The ability to muster conventional superiority, at least for a short time, against targets valuable enough to the challenger makes a strong and credible threat of conventional retaliation. The general conventional preponderance enjoyed by the challenger in all but one case does little to change the effectiveness of conventional retaliation, though it probably could if that general preponderance was transferred to change the deterrer's preponderance in the respective theater, as demonstrated by U.S. massive reinforcements with counterbattery radars in 1994.

Interestingly, in the two cases of successful deterrence, the threat of conventional retaliation was aimed at the challenger's allies. Robert Harkavy offers an early study of this deterrence strategy which he labels triangular deterrence.[9] He argues that a "new dimension of nuclear strategy and deterrence theory appear to have been heralded by the Scud attacks in 1991."[10] Working with the same empirical evidence, Kevin Wesley confirms Harkavy's ideas about the existence of triangular deterrence and argues that it would be a formidable strategy for rogue states vis-à-vis the United States.[11] Yet both Harkavy and Wesley work with the rather weak empirical cases of the Israel–Arab–Soviet triangle in the 1970s and the Gulf War SCUD attacks on Israel. In the first case, it is hard to argue with confidence that the Soviet Union was even contemplating intervention against Israel. In the second case, Iraq's threat to Israel was ineffective. My evidence offers much stronger support for triangular deterrence. The threat of China's action against U.S. allies in Asia – like South Korea, South Vietnam, and Taiwan, or even against countries such as Laos, Burma, Cambodia, Thailand, or India which did not even qualify as formal U.S. allies but which were important for the U.S. Cold War strategy – influenced Washington's decisions when a strike against Beijing's emerging nuclear arsenal was contemplated in the 1960s. Similarly, in 1994, the DPRK could not threaten more than limited retaliation against the United States, their servicemen, and their citizens. Yet, since the DPRK's threat vis-à-vis the South Korean ally was robust and credible, Washington paid attention.

In addition to challenging the analytical primacy of nuclear deterrence, my study also offers powerful inferences about nuclear deterrence with a small nuclear arsenal. Contrary to the expectation of existentialists and Waltzian optimists, nuclear posture does matter. While the two groups may be right that uncertainty always exists in deterrence relations,[12] the evidence suggests that if challengers can feel reasonably certain about the success of their first strike, they may not be deterred by a residual and highly unlikely threat of nuclear retaliation. Small, rudimentary, and largely primitive nuclear arsenals do not constitute a threat of comparable robustness to big and sophisticated arsenals. In three out of the five cases, probably only the accompanying conventional threats saved the small or emerging arsenals from being destroyed by enemy airstrike. There is little to no evidence that Washington was concerned by PRC's nuclear retaliation in the 1960s even though China could have obtained a sample bomb from the Soviet Union, could be more advanced than expected and thus armed with an indigenous bomb, or could be able to build a bomb and retaliate in the future, as proponents of post-existential deterrence like Tom Sauer would suggest.[13] The same is true for the more-than-even chance that the DPRK already had one or two bombs in 1994 and, even more importantly, for the USSR–China conflict in 1969. Back then – it appears – Moscow was largely confident that the PRC's nuclear forces could be caught on the ground and destroyed with little concern about nuclear retaliation. At the end of the day, uncertainty about a 100 percent success rate in the first strike could be largely offset by the great certainty that a similar success rate could be achieved by Soviet air defense against China's possible surviving bombers. It is also significant that when the Iraqi forces were tied in a bloody war with Iran (i.e., the accompanying conventional threat was not available), its nuclear program received a strong blow from Israel.

However, nuclear deterrence worked in one case. In the Cuban Missile Crisis the United States enjoyed a strong nuclear asymmetry by a factor of approximately 10 in strategic weapons. The Soviet Union's ICBMs were vulnerable and scarce. So were its SLBMs. Bombers were available in large numbers but also vulnerable, furthermore survivors would have to face a formidable U.S. air defense. It was uncertain that if the Americans preempted, something in the Soviet reaction (from having at least some forces surviving through communicating the decision to retaliate to penetrating the enemy's defense) would not go wrong. But, given the numbers, there was a good chance that at least some nuclear retaliation would be delivered. Plus, there were sub-strategic forces. If the Americans started a war, it would have been a nuclear one – perhaps small, but probably nuclear. At that point, the Kennedy administration did not even discuss the possibility of an all-out first strike. Despite the limits on inferences that can be drawn from one control case it can be well argued that the qualitative difference in asymmetry matters.

What theoretical observations can be made using this evidence? First, the concept of second strike has been rightly considered critical in establishing workable nuclear deterrence. However, not in the "catch all" way Kenneth Waltz suggested. Empirically, arguing that with "even a rudimentary nuclear capability, one's own severe punishment becomes possible" is largely spelling out a highly unlikely option which would only be possible if the attacker's action went terribly wrong and the defender's action passed flawlessly.[14] In fact, there is a difference between having a rudimentary nuclear arsenal and having a second-strike capability. True, nuclear weapons can be hidden and thus difficult to locate for the possible attacker. But that does not ensure that they can be delivered against targets valuable for the attacker. Waltz's expectation that delivery vehicles do not matter is not grounded in evidence. Nothing supports the estimate that nuclear weapons can be detonated by trucks or boats lying offshore. Or more accurately they can, but no country has ever entrusted its deterrence to trucks or boats and no country has ever been afraid of that. These were allegedly the only delivery options for North Korea's two possibly existing bombs in 1994. No precautions were taken to deal with such a threat, though apparently stopping trucks from crossing the border that normally no one crosses would not be too difficult. The U.S. decision makers did not express any concerns about the DPRK's nuclear-armed trucks.

My second theoretical observation is that proliferation optimism stands largely unwarranted. While nuclear weapons may constitute a robust and credible deterrence, they are neither easy nor cheap. Robust deterrence requires second-strike ability, which is more demanding than the mere possession of some nuclear weapons that may or may not survive a surprise attack. Furthermore, emerging proliferators would likely have to go through the dangerous window of vulnerability when nuclear weapons do not deter more than they upset others and invite interventions. The time from detection of proliferation to achievement of a real second-strike ability can be precariously long. Yet empirical evidence suggests that proliferation is likely to cause states that are already hostile to the proliferator to contemplate prevention. While optimists may counter that nuclear proliferation has only rarely led to prevention, empirical evidence strongly suggests that the lack of interventions had nothing to do with the stabilizing effects of proliferation. There is little reason to hail proliferation for its stabilizing effects when in reality what deterred were conventional arms. Proliferation likely offers little stability, and certainly not easy stability.

Unlike proliferation optimism, two other schools probably describe deterrence well or, more accurately, give correct prescriptions for policy. First, minimalists argue that asymmetry does not matter much as long as the second strike criterion is fulfilled – this is more in line with Wohlstetter than Waltz. At least this conclusion can be drawn from the single case of successful nuclear deterrence in this study. Despite the strong asymmetry between the United States and the Soviet Union in 1962, the Soviet nuclear arsenal strongly inhibited the U.S. option to escalate. It appeared clear to Kennedy and his top advisors that should the conflict escalate the war would be nuclear. Yet, clearly, the Soviet ability to damage the United States was far from the requirements of mutual assured destruction. A stronger and more sophisticated Soviet arsenal would have been able to bring more destruction upon the United States, but the improvement in the robustness of Soviet nuclear deterrence would have been marginal. The same is likely not true of the U.S. arsenal. This, perhaps surprisingly, makes the claims of war-fighters also compatible with the empirical evidence. Had the United States enjoyed the necessary qualitative and quantitative superiority to break the Soviet ability to strike back, they would have also enjoyed the freedom to act which supporters of war-fighting desire. Thus since the second-strike criterion is not easy to achieve – and nuclear deterrence does not appear to be effective unless the criterion is met – asymmetry can matter. While minimalists suggest that it is best to stay on the side of having the second-strike ability to deter, proponents of war suggest it is best to be able to beat the other's second strike in order not to be deterred. The same empirical evidence can confirm both opinions because the minimalists in fact offer a deterrence strategy, while the proponents of war-fighting, in reality, propose a counter-deterrence strategy.

Tentative findings

A number of additional interesting and important observations can also be made with respect to other concepts that together comprise the framework of this research. Admittedly, some of those observations are constrained by the limits of this research – such as the number and character of cases and the imperfect evidence available on them. I am the first to admit that such results are only tentative, yet, I would argue, valuable for further refinement.

First, it appears that deterrence with a small nuclear arsenal is particularly prone to failure when the small arsenal is vulnerable to a conventional attack. In all cases, the challengers showed remarkable preference for conventional weapons, while a nuclear first strike appeared rather unattractive to them. Surprisingly, while this is compatible with the expectations about normative prohibition of nuclear first use, it also seems that the evidence does not strongly support the normative logic of nuclear taboo. The challengers' preferences for conventional weapons more likely followed the logic of consequences rather than the logic of appropriateness. In 1963, the JCS recommended tactical nuclear weapons for a possible attack against China's nuclear installations, yet the administration overruled them and ordered the military to plan a conventional operation, expecting to moderate China's reaction and reduce political costs. In 1969, most Soviet Politburo members preferred conventional options, but mostly because they believed they were sufficient. In 1994, the Clinton administration did not even discuss the first strike option, yet the value of the evidence for a nuclear taboo is unclear. Modern conventional weapons were militarily sufficient; nuclear weapons would have brought little benefits for the mission. This evidence is largely in line with recent empirical research by Press, Sagan, and Valentino on public attitudes to using nuclear weapons, which suggests strong support for the fact that those attitudes are driven largely by consequentialist considerations of military utility.[15] By choosing the conventional option for taking out the deterrer's nuclear arsenal, the challengers usually expected reduced political costs, improved legitimacy of action, and also, importantly, leaving the burden of escalation to the nuclear level to the deterrer. Interestingly, this preference for conventional weapons appears to exist across various regime types. The democratic government in the United States ordered its military to plan a conventional strike against China's nuclear installations despite the recommendation of its top military advisors to use tactical nuclear weapons. The Soviet Politburo preferred a conventional strike against the suggestion of its highest-ranking soldier, Marshal Grechko, to deal with the Chinese problem once and for all with nuclear weapons.

With likely improvements of conventional weapons, possible future challengers will probably find it even easier to deny their nuclear-armed opponents second-strike ability. Today's conventional arms have such destructive power against point targets that in many respects they surpass the counterforce abilities of Cold War nuclear weapons. Preventive strikes can now be cheaper than before. In 1963, the conventional destruction of China's nuclear installations would have required a large number of sorties and corresponding losses of attacking aircraft. In 1981, Israel could destroy Osiraq with 16 conventional bombs which were dropped from eight F-16s in a high-risk yet casualty-free mission, and, in 1994, the Americans could easily destroy Yongbyon with cruise missiles and stealth bombers in a mission with little direct danger for U.S. soldiers. This development undermines proliferation optimists' scholarship even further. States are often tempted to stop proliferation during the window of vulnerability. Modern conventional arms give a would-be attacker preferable tools for striking, and due to technical improvements, they also make the window of vulnerability larger and longer-lasting.

Other factors look less decisive. There was a strong technological advantage on the part of the challenger in all but the control case. Yet technological advantage alone apparently guarantees little. The United States enjoyed a strong advantage over China in the early 1960s, yet the Korean War had taught them that China's advantage in manpower could not be easily broken by advantages in technology. Similarly, in 1969, the Soviets had more advanced weapons in the Far East than the Chinese, but not enough to change the fact that they were outnumbered by more than two to one in terms of personnel. The revolutionary American conventional weapons would certainly have prevailed against North Korea in 1994, but not until they were deployed in the theater. The DPRK's conventional preponderance would have likely been short-lived, but not short enough to save Seoul. Only when technological advantage is transferred into theater conventional preponderance will technology be decisive.

The role of the availability of information is not completely clear. In the comparative perspective, the strong availability of information occurred in the single case of deterrence failure. Intuitively such availability should be a necessary but not a sufficient condition. The latter is certainly true. The Soviet Union likely had enough information about China's nuclear arsenal yet it did not strike. In the other three cases, the challenger had only limited information available and refrained from attacking. However, it is uncertain whether the lack of information made the strike impossible. Apart from the U.S.–DPRK case, decision makers were not too concerned about the inadequacy of targeting intelligence and, even in that case, the strike might have been possible without complete intelligence including the location of the plutonium that had previously been extracted from the reactor.

In all cases, the conflict eventually appeared central to the challenger. But only limited conclusions can be drawn there. Certainly it can be stated that even in the central conflicts deterrence may hold as well as fail. Yet this is not a groundbreaking observation. Observing the cases, it can also be stated that a relation exists between the centrality of conflict and general deterrence failure in the nuclear dyads. Yet it is impossible to distinguish whether general deterrence fails because the conflict is central, or that it becomes central when general deterrence fails. Most profoundly, in the U.S.–DPRK case, the conflict became truly central only after the failure of general deterrence. As the conflict escalated, the Clinton administration had no other choice than to put it on the top of its list of serious international crises. On the other hand, in the U.S.–China case, general deterrence largely failed because the conflict was central. The Kennedy administration would have not contemplated striking the PRC's emerging nuclear arsenal unless the conflict with China was central to it. General deterrence failure may precede centrality of conflict, or arise from it. Only further research can give more details.

Interesting relations also likely exist between the perception of resolve and the perception of rationality. Apart from the control case, the deterrer's rationality was strongly questioned. Lack of rationality is a serious problem for mainstream deterrence theory which assumes both challenger and deterrer are rational. Yet it also helps solve one of the theory's problems. For a rational deterrer, responding to an attack may be irrational. But fools are believed to retaliate even if it brings them little benefit.[16] Empirically, it can be observed that the alleged lack of deterrer rationality also contributed to the perception of his resolve. Mao's references to nuclear weapons made leaders in both Washington and Moscow gravely concerned about China's rationality, and, at the same time, they did not question whether China would respond to an attack. On the contrary, in the control case the deterrer's rationality was the least questioned but his resolve was put in question by the proponents of the air strike, who argued that the Kremlin would not risk its own destruction by escalating the crisis should the missiles in Cuba be destroyed. Their argument appeared to be overruled on the grounds that such an irrational decision could be taken lower in the Soviet chain of command. None of this should be surprising for a careful student of deterrence and a reader of Thomas Schelling.[17] However, that does not mean that being perceived as irrational helps stability. The increased perception of resolve is outweighed by the general distaste for living with a nuclear-armed irrational actor. Though not necessarily, this may lead to last-resort considerations which greatly strengthen the support for preventive options. Certainly this logic appeared in Kennedy's thinking about China and Begin's about Iraq, in the latter case leading to an actual pre-emption as no robust threat was available to dissuade Israel's military action.

The other two perceptional factors look less critical. Deterrence successes occurred in cases with a strong history of hostility such as those between the United States and the PRC or the USSR in the 1960s, as well as in the Sino-Soviet case where the history of hostility was rather limited. The same is true for the institutionalization of mutual relations. Success came with strong institutionalization in the Sino-Soviet case, but also with limited institutionalization in the U.S.–DPRK case (where Sigal observes that for the Americans, North Korea was such an unknown incomprehensible country that it could have been on Mars with little difference).[18]

What about international factors that have not been addressed yet? There is not much to say about third-party deterrence. It played nothing more than a limited role in any of the five cases. Surprisingly strong concerns about alliance politics occurred in the single case of deterrence failure in this study as well as in the two cases of successful deterrence. In 1962, the Kennedy administration felt that U.S. allies would find the air strike against the missiles in Cuba unacceptable. In 1994, the Clinton administration paid enormous attention to South Korea's and Japan's reactions to an action against North Korea and slowly built a consensus for such an action. In both cases, deterrence worked. Yet Israeli concerns about possible damage that Osiraq's destruction would cause to the country's vital alliance with the U.S. were hardly smaller. Obviously, while alliance politics matters it is not necessarily prohibitive enough. It is also notable that stronger concerns about international action legitimacy appeared in the case of deterrence failure than in some cases of deterrence success. Apparently, in 1981, Israel was fully aware that its action would be troublesome from the perspective of international norms. The concerns were vocally voiced during the cabinet's meetings, yet the Iraqi reactor ended in the rubble. On the contrary, in 1963, Kennedy's administration raised concerns about action legitimacy, but did not consider it prohibitive. The Soviets in 1969 likely cared even less. By 1994, Pyongyang was believed to be in serious breach of international norms and acting against it would have been largely legitimate. However, in contrast to the Israel–Iraq case, the strike did not occur. Again it is reasonable to say that a lack international action legitimacy is not necessarily prohibitive.

The same cannot be observed with regard to domestic action legitimacy. In four cases, a hostile action against enemy's nuclear weapons appeared largely legitimate. In neither case of a democratic challenger did the domestic audience show a strong inclination to keep the peace. While the evidence is only tentative, it largely corresponds to the argument that democracies may not fight each other, but are hardly peaceful in general.[19] In fact, the domestic audience appears to be often more hawkish then the respective government. In one case (USSR–PRC) domestic action legitimacy is hard to judge but apparently did not play much of a role in Soviet decision makers' considerations. Again, this is a largely expected observation with respect to the type of regime in the USSR. But, in general, regime type does not strongly influence the ways deterrence works or fails. In fact, regime type effects on most of the factors in the study seem to be marginal.

Further research and policy recommendations

My research shows that a lot needs to be re-examined and re-conceptualized in what is considered today to be common knowledge about the relationship between conventional and nuclear deterrence and the effectiveness of deterrence with small nuclear arsenals. There are two fundamental biases students of deterrence should consciously avoid: the empirically unwarranted analytical primacy of nuclear deterrence, and the existential bias. Furthermore, more empirical research is needed in three other areas.

First, thorough attention should be paid to the way conventional and nuclear deterrence operate side by side. So far, this area has been gravely understudied. Researchers study nuclear deterrence in nuclear dyads and conventional deterrence in conventional dyads, or alternatively in stability-instability paradox driven situations in nuclear dyads. This is insufficient. Deterrence in nuclear dyads should be re-examined to better highlight the role played by conventional deterrence. Obviously, historical case studies would be by far the most suitable method for this research.

Second, further research is needed with respect to conventional retaliation. Existing literature treats conventional deterrence as denial-dominated and conventional retaliation as ineffective. In fact, the concept itself is far from being sufficiently refined. Huntington's rare article offers useful guidance, but it aims to formulate propositions for the specific policy problem of the conventional balance in 1980s, Cold War Europe, rather than as a full-fledged theoretical development of the concept. Thus, conventional retaliation desperately needs both careful theoretical refinement and extensive empirical research.

Third, empirical research will also be useful with respect to refining my tentative findings. Is the likely development of even more sophisticated conventional weapons going to increase incentives for preventive strikes against emerging nuclear arsenals? What role does the consequential logic of military utility play in the nuclear taboo? Historically, how effective was triangular deterrence? Why does regime type appear to have so little influence in deterrence? Those are only some important questions to be addressed. As with most research, this study answers some questions, but leaves many unanswered and opens yet others.

Further research in the directions outlined above would allow formulating stronger policy recommendations that I can give now. At this point, aware of the limitations of my research and the significance of the topic, I would limit my practical suggestions to the preliminary level. However, that does not mean they are unimportant.

Most importantly, states should be cautioned against proliferation. The most likely candidates for future proliferation – desperate nations with strong hostile neighbors and without viable alliances – are also the most likely to put themselves in great danger by pursuing nuclear options. Instead of the desired stability, the proliferator could get instability or even war. Nations without a strong conventional army could endanger themselves most because they would lack the protective shield of conventional deterrence that saved the three deterrers in this study from foreign intervention. To give one example, today's Ukraine is a pre-eminent candidate that should be cautioned. Strong countries do not like it when their adversaries seek nuclear arms, and Ukraine lives next to Russia. Moscow may contemplate prevention, or even opt for it. It would not be the first great power to be tempted.

This book has also highlighted the stronger effectiveness of conventional deterrence than is generally believed. Accordingly, it could also be suggested that states should be encouraged to rely more on those means. Probably, nuclear weapons can be replaced with conventional weapons for deterrence strategies against most threats. Until more research in this direction confirms my results, I would be careful to suggest replacing nuclear weapons in all their functions. Perhaps they should be kept for deterring a nuclear attack until verifiable nuclear disarmament can be achieved. Yet, for most other deterrence missions, nuclear weapons are likely to be replaceable with proper conventional strategies, and a no-first-use policy can be safely adopted.

Notes

1 Patrick M. Morgan, *Deterrence: A Conceptual Analysis* (Sage: London 1977) 58–59.
2 Edward Rhodes, "Conventional Deterrence," *Comparative Strategy*, 19/1 (2000) 221–253.
3 Richard K. Betts, "Conventional Deterrence: Predictive Uncertainty and Policy Confidence," *World Politics*, 37/2 (January 1985) 177.
4 Leon V.Sigal, *Disarming Strangers: Nuclear Diplomacy with North Korea* (Princeton: Princeton University Press, 1999) 122.

5 John J. Mearsheimer, *Conventional Deterrence* (Ithaca: Cornell University Press, 1983) 23–24.

6 Rhodes, "Conventional Deterrence."

7 Robert Pape, *Bombing to Win: Air Power and Coercion in War* (Ithaca: Cornell University Press, 1996) 21–27.

8 Mearsheimer, *Conventional Deterrence*, 24–28; also see the substantial body of literature on offense-defense balance. A good overview is available in Sean M. Lynn-Jones, "Offense-Defense Theory and Its Critics," *Security Studies*, 4/4 (Summer 1995) 660–691; Stephen Biddle, "Rebuilding the Foundations of Offense-Defense Theory," *The Journal of Politics*, 63/3 (August, 2001) 741–774; and Michael E. Brown, Owen R. Coté, Sean M. Lynn-Jones, and Stephen E. Miller (eds), *Offense, Defense, and War* (Cambridge: MIT Press, 2004).

9 Robert Harkavy, "Triangular or Indirect Deterrence/Compellence: Something New in Deterrence Theory?," *Comparative Strategy*, 17/1 (1998) 63–81.

10 Harkavy, "Triangular or indirect deterrence/compellence," 77.

11 Kevin R. Wesley. *Triangular Deterrence: A Formidable Rogue State Strategy.* Monterey, 1999. Masters Thesis. Naval Postgraduate School.

12 Kenneth N.Waltz, "More May Be Better," in Scott D. Sagan and Kenneth N. Waltz, *The Spread of Nuclear Weapons: A Debate Renewed* (New York: W.W. Norton, 2003) 3–45; McGeorge Bundy, "Existential Deterrence and its Consequences" in Douglas MacLean (ed.), *The Security Gamble: Deterrence Dilemmas in the Nuclear Age* (Totowa: Rowman and Littlefield, 1984) 3–13.

13 Tom Sauer, "A Second Nuclear Revolution: From Nuclear Primacy to Post-Existential Deterrence," *Journal of Strategic Studies*, 32/5 (October 2009) 745–767.

14 Waltz, "More May Be Better," 19.

15 Daryl G. Press, Scott D. Sagan, and Benjamin A. Valentino, "Atomic Aversion: Experimental Evidence on Taboos, Traditions, and the Non-Use of Nuclear Weapons, *American Political Science Review*, 107/1 (February 2013) 188–206.

16 Patrick M. Morgan, *Deterrence Now* (West Nyack: Cambridge University Press, 2003), Chapter 2.

17 Thomas C. Schelling, *Arms and Influence* (New Haven: Yale University Press, 1966) 36–43.

18 Sigal, *Disarming Strangers*, 10.

19 See Anna Geis, Lothar Brock, and Harald Müller, *Democratic Wars: Looking at the Dark Side of Democratic Peace* (New York: Palgrave Macmillan, 2006).

Index

Taylor & Francis eBooks

Helping you to choose the right eBooks for your Library

Add Routledge titles to your library's digital collection today. Taylor and Francis ebooks contains over 50,000 titles in the Humanities, Social Sciences, Behavioural Sciences, Built Environment and Law.

Choose from a range of subject packages or create your own!

Benefits for you

>> Free MARC records
>> COUNTER-compliant usage statistics
>> Flexible purchase and pricing options
>> All titles DRM-free.

REQUEST YOUR FREE INSTITUTIONAL TRIAL TODAY

Free Trials Available
We offer free trials to qualifying academic, corporate and government customers.

Benefits for your user

>> Off-site, anytime access via Athens or referring URL
>> Print or copy pages or chapters
>> Full content search
>> Bookmark, highlight and annotate text
>> Access to thousands of pages of quality research at the click of a button.

eCollections – Choose from over 30 subject eCollections, including:

Archaeology	Language Learning
Architecture	Law
Asian Studies	Literature
Business & Management	Media & Communication
Classical Studies	Middle East Studies
Construction	Music
Creative & Media Arts	Philosophy
Criminology & Criminal Justice	Planning
Economics	Politics
Education	Psychology & Mental Health
Energy	Religion
Engineering	Security
English Language & Linguistics	Social Work
Environment & Sustainability	Sociology
Geography	Sport
Health Studies	Theatre & Performance
History	Tourism, Hospitality & Events

For more information, pricing enquiries or to order a free trial, please contact your local sales team:
www.tandfebooks.com/page/sales

 **Routledge**
Taylor & Francis Group

The home of
Routledge books

www.tandfebooks.com